Practical Electronics Crash Course

Learning circuit design the fun way

Dr. Dogan Ibrahim
Ahmet Ibrahim

elektor
design > share > earn

This is an Elektor Publication. Elektor is the media brand of
Elektor International Media B.V.
PO Box 11, NL-6114-ZG Susteren, The Netherlands
Phone: +31 46 4389444

British Library Cataloguing in Publication Data
A catalogue record for this book is available from the British Library

ISBN 978-3-89576-605-3 Print
ISBN 978-3-89576-606-0 eBook

www.elektor.com
Editor: Clemens Valens
Prepress Production: D-Vision, Julian van den Berg
Printers: Ipskamp, Enschede, The Netherlands

Contents

Preface

Welcome to the world of electronics!
In this beginner-friendly book, you will explore and learn the most important electrical engineering and electronics concepts in a fun way by doing various experiments and by simulating circuits. The aim of the book is to teach you the basic electronics without getting into complex technical jargon and calculations.

Learning electronics is not as difficult as it may seem. Using this book, you will be creating your own projects sooner than you might think. The book starts from the very basics, such as what is electricity? What is voltage? What is current? and so on. You will then learn about simple electronic circuits and the symbols used in circuits. No previous knowledge of electronics is required, only some simple algebra is required to be able to make simple calculations. Circuit simulation has been introduced at an early stage to enable readers to experiment with different circuits.

You will learn:

- The concepts of voltage, current, circuit, and power
- AC and DC voltages
- Basic lamp circuits with switches
- Resistors, capacitors, inductors and how to identify them
- Transient circuits
- Electromagnetism
- Loudspeakers, reed switches, and transformers
- Diodes, BJT and MOSFET transistors and switching circuits
- Optocoupler circuits
- Multivibrator and astable/monostable circuits
- Using the 555 chip
- Operational amplifiers and various circuit configurations
- Logic gates
- Amplifiers, oscillators, and electrical filters
- Sensors
- Electrical test and measurement tools
- Microcontroller development boards: Arduino Uno, ESP32, Raspberry Pi Pico, and Raspberry Pi 5
- Electronic component selection and reading data sheets
- Electrical EMC and EMI and norms and regulations

Many tested and working projects and simulations are given in the book to make you familiar with the construction and testing of the basic electronic circuits given in the book. I hope you enjoy reading the book.

Dr. Dogan Ibrahim, BSc., MSc., PhD
Ahmet Ibrahim, BSc., MSc
London, 2024

Chapter 1 • Electricity

1.1 Overview

We all use electricity in our daily lives, and we cannot do without it. Can you imagine what will happen if electricity was not available for a few days? Imagine life with no heating, no lighting, no radio, no TV, no internet, no computer, no microwave, no fridge... But have you ever asked yourself: What is electricity? What is voltage? What is an electrical circuit? How can I connect lamps in different ways? The aim of this chapter is to answer these questions in detail.

1.2 What is Electricity?

Atoms are the building blocks of matter; they are the smallest units of matter that have the chemical properties of a given substance. Atoms comprise three basic components: protons, neutrons, and electrons (Figure 1.1). The protons and neutrons are in the center of the atom, known as the nucleus, while the electrons orbit at a considerable distance from the nucleus.

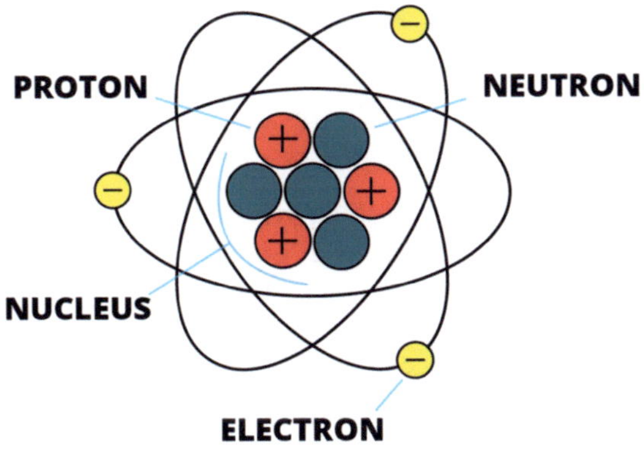

Figure 1.1 An atom. (source: www.vedantu.com)

The atomic number of an atom corresponds to the number of protons in the nucleus. For example, a carbon atom has six protons in its nucleus, therefore its atomic number is six. The sum of the number of protons and neutrons determine the atomic mass of an atom.

Protons are positively charged particles inside the nucleus. The neutrons inside the nucleus do not have charge. The mass of a neutron is more than that of a proton. Electrons are negatively charged particles that orbit the nucleus within orbital shells. They are much smaller than protons and neutrons. The overall charge of an atom is determined by the difference between the number of protons and electrons. For example, if there are more protons than electrons, the net charge is positive. If the number of protons is equal to the number of electrons, the net charge is zero.

Electricity is the set of physical phenomena associated with the presence and motion of electric charge. Electrons can flow only if there are free electrons in a substance. Concerning the flow of electrons, materials fall into three categories:

Insulators: Some materials like wood, paper, plastic, rubber, etc. do not conduct electricity and therefore are good insulators. This is due to the (very) low number of free electrons that are available.

Conductors: Many metals such as copper, iron, silver, steel, etc. are excellent conductors of electricity. This is because they have an abundance of free electrons.

Semiconductors: Materials like germanium, arsenic, boron, and silicon have only a few free electrons. They conduct electricity better than insulators but not as good as conductors do. Semiconductors are used for manufacturing transistors and integrated circuits (ICs).

1.3 Electronic Circuits

For electrons to flow, an electronic (or electrical) circuit is required. A circuit is a closed loop of conducting material such as wire. Of course, every electronic circuit also contains components, but there may not be insulators in the circuit loop. The electrons in an electronic circuit do not move unless they are forced to do so. This force can be a battery or any other type of energy source connected to the circuit. The movement of electrons in a circuit constitutes the current in the circuit. Thus, when a voltage source is applied to a circuit, current can flow through it. The electrons flow from the negative pole towards the positive pole. But, because electrons have a negative charge, it is customary to say that the current flows from plus to minus.

The unit of voltage is volt, denoted by V (sometimes also denoted by U). The unit of current is ampere, A. In most electronic applications, smaller units of volt and ampere may be used. The conversions to smaller units are as follows:

1 V = 1,000 millivolt (1000 mV)
1 mV = 1,000 microvolt (1000 μV)

Similarly,

1 A = 1,000 milliampere (1000 mA)
1 mA = 1,000 microampere (1000 μA)

1.4 DC and AC

There are two types of voltage sources: Direct Current or DC and Alternating Current or AC.

DC

In a DC circuit, the current always flows in the same direction. A DC voltage source does not necessarily have a constant value, but it never changes sign. An example of a DC voltage source is an ideal battery, which provides a constant voltage over time. For example, an ideal PP-3 type battery provides 9 V as shown in Figure 1.2. This is described mathematically as:

$$V_{(t)} = 9\ \text{V}$$

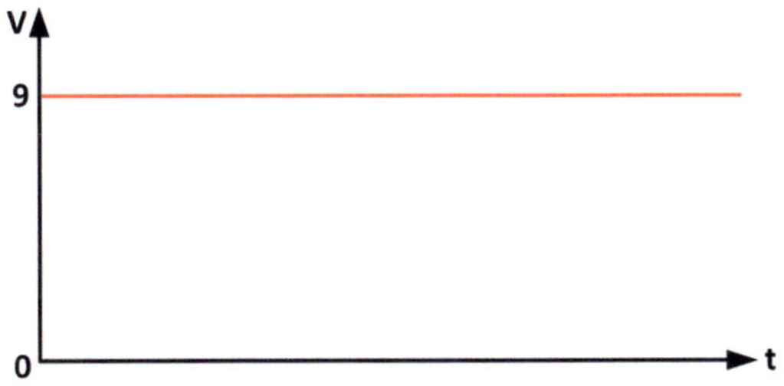

Figure 1.2 9V battery voltage.

Of course, a battery loses its charge over time and its voltage drops as the battery is used, but for the discussion here, we assume that the voltage is constant.

AC

In an AC circuit, the direction of the charge flow and hence the voltage level reverses periodically. Electricity providers supply AC voltages known as the mains voltage to buildings all over the country. However, most electronic devices like mobile phones, computers, radios, TVs, etc. operate with DC voltages and the AC voltage must be converted into DC. As an example, your mobile phone has a battery and operates from a DC voltage, but you plug its charger into an AC wall socket. The charger converts the AC to DC to charge your phone's battery.

AC voltages can have many shapes like sine, square, triangle, sawtooth, etc. The AC voltage supplied by electricity providers is in the form of a sine wave (Figure 1.3), expressed mathematically as:

$$V_{(t)} = V_p \sin(\omega t)$$

where

$V_{(t)}$ is the voltage as a function of time.
V_p is the amplitude (peak value) of the waveform. The waveform spans from $-Vp$ to $+Vp$.
ω is the radian frequency equal to $2\pi f$, where f is the frequency of the waveform in hertz.
t is the time in seconds.

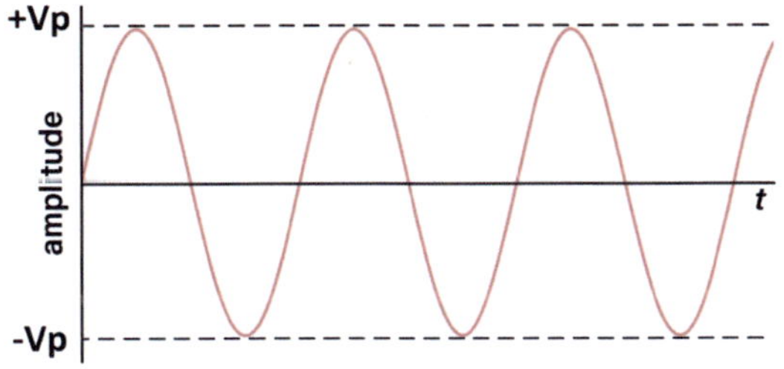

Figure 1.3 Sine waveform.

The period of a sine wave is the time it takes for a cycle to complete and is related to the frequency by the relationship

$$T = 1/f$$

Example
Find the period and hence the frequency of the sine wave shown in Figure 1.4. Write down the mathematical expression of the waveform.

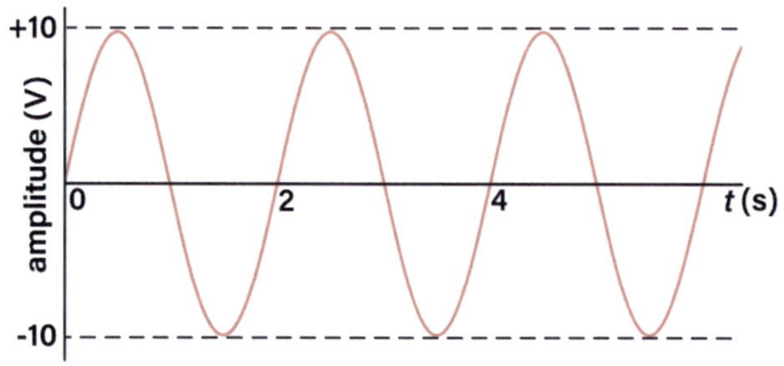

Figure 1.4 Example sine wave.

Solution
The amplitude is 10 V. The period of the waveform is found to be T = 2 seconds. Therefore, the frequency is f = 1/2 = 0.5 Hz. The radian frequency is, $\omega = 2\pi f = 2 \times 3.14 \times 0.5 = 3.14$ rad/s. The mathematical expression of the waveform therefore is:

$$V_{(t)} = 10 \sin(3.14t)$$

1.5 Electric Power

Electric power is the rate at which electrical energy is transferred by an electric circuit. In other words, it is the rate of work done per unit of time by an electric circuit. The unit of electric power (P) is watt (W), given by:

$$P = \text{current} \times \text{voltage}$$

Where power is in watt, current is in ampere, and voltage is in volt. Power is sometimes expressed in bigger or smaller units, as follows:

1 kW = 1,000 watt (1000 W)
1 W = 1,000 milliwatt (1000 mW)
1 mW = 1,000 microwatt (1000 μW)

Power can also be calculated from the energy in joule over a period of time. i.e.

P in W = energy/second = joule/second

Example

A mains operated kettle works with 230 V and consumes 2 A. Find the power consumed by this kettle. If the kettle is operated for 10 seconds, how much electrical energy is used?

Solution

From the above equation, $P = 230 \times 2 = 460$ W.
The energy used is $E = P \times t = 460 \times 10 = 4{,}600$ joules.

1.6 Basic Lamp/Switch Circuits

Suppose you have some lamps and a few switches. Let us investigate in how many ways you can connect them in a circuit. In the following circuits, V_{Ln} is the voltage across lamp *n*, and ILn is the current through lamp *n*. Also, it is assumed that the brightness of the lamps is the same as long as current passes through them.

1.6.1 One Switch Only

Series connection of lamps

Figure 1.5 shows a series connection of three lamps and a switch. Here, the same current goes through each lamp. The sum of the voltages across each lamp is equal to the applied voltage. When the switch is closed, all lamps turn on.

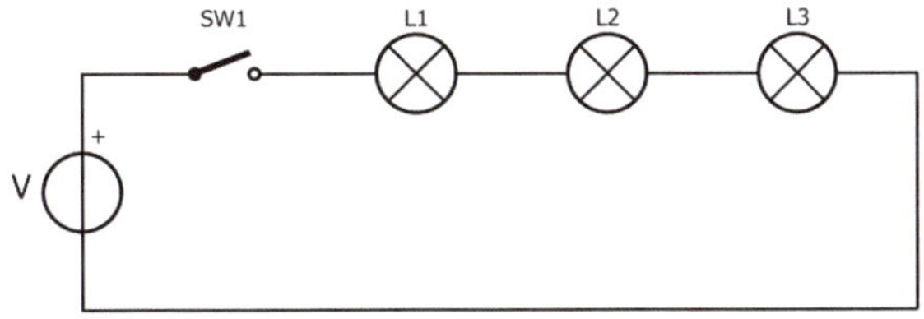

Figure 1.5 Series connection. The same current flows through all the lamps. When the switch is closed, all lamps turn on. $V = V_{L1} + V_{L2} + V_{L3}$

Parallel connection of lamps

Figure 1.6 shows a parallel connection of three lamps and a switch. Here, the same voltage is across each lamp. The sum of the currents through the lamps is equal to the main current. When the switch is closed, all lamps turn on.

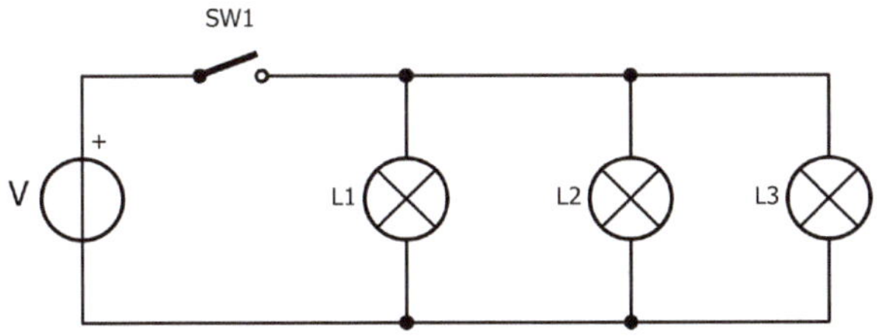

Figure 1.6 Parallel connection. Each lamp has the same voltage across it. When the switch is closed, all lamps turn on. $I = I_{L1} + I_{L2} + I_{L3}$

Series and parallel connection of lamps

Figure 1.7 shows a mixed series and parallel connection of three lamps and a switch. Here, the same voltage is across L2 and L3. The sum of voltages across L1 and L2 (or L3) is equal to the applied voltage. The current through L1 is equal to the sum of the currents through L2 and L3.

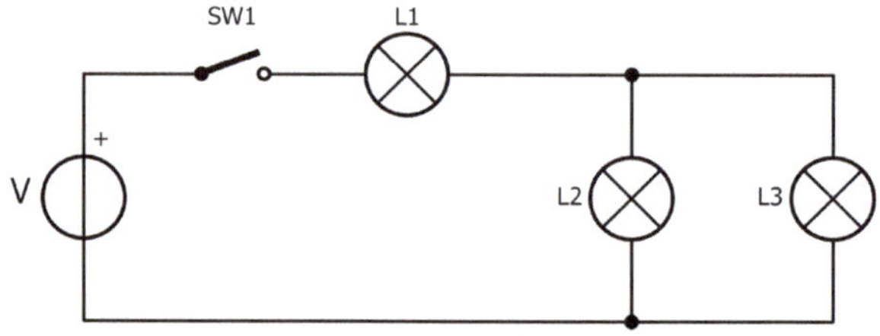

Figure 1.7 Series and parallel connection.
Lamps L2 and L3 have the same voltage across them.
When the switch is closed, all lamps turn on. $V = V_{L1} + V_{L2}$ *or* $V = V_{L1} + V_{L3}$; $I_{L1} = I_{L2} + I_{L3}$

1.6.2 More Than One Switch

Figures 1.8 to 1.10 show various configurations of lamps and switches. The operation of each circuit is described below the circuits.

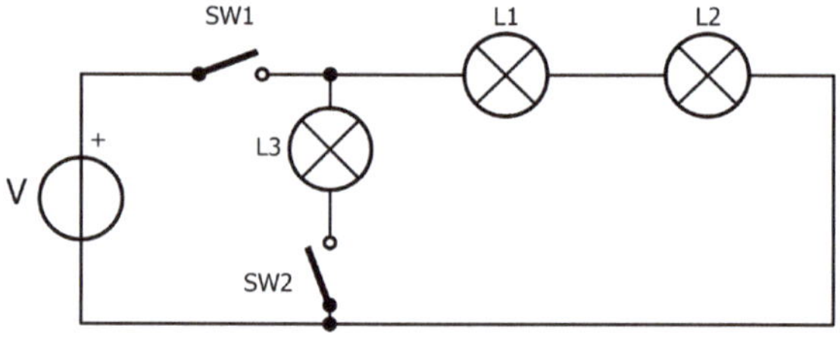

Figure 1.8 L1 and L2 turn on when SW1 is closed. L3 turns on when SW2 is closed.

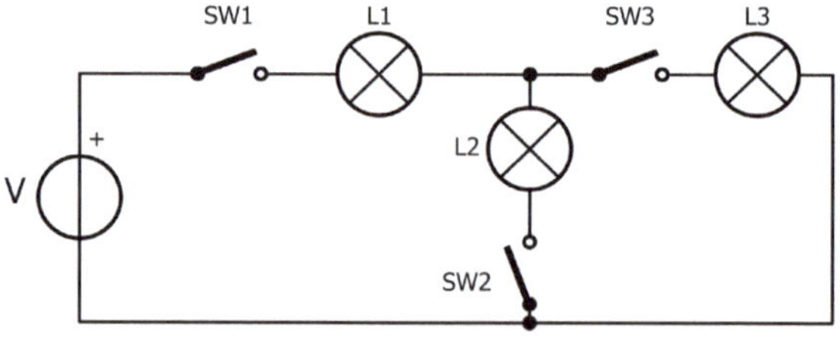

Figure 1.9 L1 turns on when SW1 is closed and also SW2 or SW3 are closed.
L2 turns on when SW1 and SW2 are closed (L1 turns on too).
L3 turns on when SW1 and SW3 are closed (L1 turns on too).
All lamps turn on when all switches are closed.

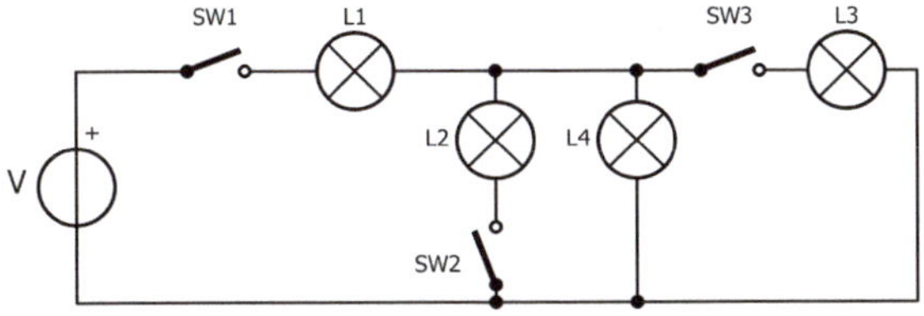

Figure 1.10 L1 and L4 turn on when SW1 is closed. L2 turns on when SW1 and SW2 are closed (L1 and L4 turn on too). L3 turns on when SW1 and SW3 are closed (L1 and L4 turn on too). All lamps turn on when all switches are closed.

1.6.3 Two-Way Switch

You are probably familiar with two-way switches if your home has a staircase or a long hallway. They are also often found in bedrooms. In a two-way switch configuration, a lamp is controlled by two switches, for example one downstairs and one upstairs. If the lamp is off, it can be turned on from the downstairs switch. Once you are upstairs, you can turn it off from the upstairs switch and vice versa. Figure 1.11 shows the schematic of a two-way switch. This circuit uses two switches of type SPDT (Single Pole Double Throw). The operation of various switches is discussed in Chapter 5.

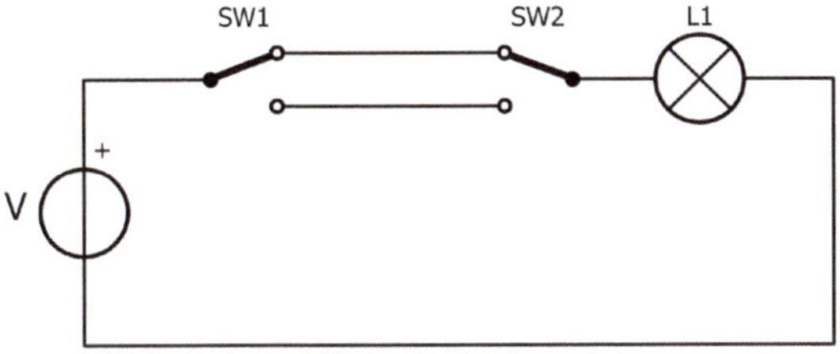

Figure 1.11 Two-way switch. When the lamp is on, you can turn it off with either SW1 or SW2. When the lamp is off, you can turn it on with either SW1 or SW2.

Chapter 2 • Passive Components

2.1 Overview

Passive components are parts like resistors, capacitors, inductors, cables, switches, fans, lamps, antennas, terminals and connectors, coils, transformers, printed circuit boards (PCBs) and so on. These components cannot generate power, but they receive electrical energy, which they can either absorb, dissipate, or store. Passive components do not provide any gain or amplification of electrical signals. They can either be used on their own or connected in series or in parallel with other components in electrical circuits to control the current flow, apply a phase shift to signals, change the voltage across a component and so on. Passive components are bidirectional (they do not have polarity) as they can be connected in either direction within a circuit (except for electrolytic capacitors as they have polarity).

In this chapter, we will be looking at the most commonly used passive components: resistors, capacitors, and inductors, and their combinations in electrical circuits.

2.2 Resistors

Resistors play a crucial role in electrical circuits by impeding the flow of electric current. To comprehend the function of a resistor more easily, one can draw an analogy between electric current and water flowing in a pipe. In this analogy, a resistor is akin to a constriction in the pipe, acting to limit the flow of water through it. The degree of restriction in the pipe correlates with the resistance in the electrical circuit. Simply put, the higher the resistance, the more constrained the flow of electric current becomes. Similarly, in the water analogy, an increased constriction in the pipe results in a reduced volume of water able to pass through.

Resistors have many functions in electrical circuits. Some common uses of resistors are:

- **Potential dividers:** two or more resistors in series are used to reduce an input voltage to a required level.
- **Current limiting:** a resistor is used to limit the current flow in an electrical circuit, usually to protect a device. For example, a current limiting resistor is used in LED circuits to limit the current and hence protect both the current source and the LED.
- **Operational amplifiers:** Here, resistors are used to set the amplifier gain.
- **Active filters:** Resistors are used in active filter circuits to set the filter gain, frequency, and phase responses.
- **Current measurement:** Resistors are used to measure the currents through circuits.

- **In active circuits:** Resistors are used in various active circuits, such as amplifiers, mixers, oscillators, comparators, buffers, transmitters, receivers, etc. to set the circuit characteristics.

Resistor values

Resistor values are in ohms, denoted with the Greek capital letter omega (Ω). For example, 100 ohms or 100 Ω. Larger values are available as kilohms, denoted as kΩ, or megohms, denotes as MΩ. 1 kΩ is equal to 1,000 Ω and 1 MΩ is equal to 1,000,000 Ω. Therefore, 1 MΩ is equal to 1,000 kΩ.

Resistor values have a tolerance, specified as a percentage. For example, the value of a 100 Ω resistor with a 10% tolerance may lie in the range from 90 Ω up to 110 Ω. Common tolerances are 10%, 5% and 1%.

Because it is complicated to produce and stock resistors of every value possible, they have been organised into series of preferred values known as the E-series. The E-series are based on tolerances. The number following the 'E' indicates the number of different values in a series. The E12 series, the most used, is based on 10% tolerance and has 12 values per decade:

1.0, 1.2, 1.5, 1.8, 2.2, 2.7, 3.3, 3.9, 4.7, 5.6, 6.8 and 8.2.

For example, within the E12 series, we can find resistors with the following values in the 1.8 range:

1.8 Ω, 18 Ω, 180 Ω, 1.8 kΩ, 18 kΩ, 180 kΩ, 1.8 MΩ and so on.

The E24 series is based on 5% tolerance and has 24 values per decade:

1.0, 1.1, 1.2, 1.3, 1.5, 1.6, 1.8, 2.0, 2.2, 2.4, 2.7, 3.0, 3.3, 3.6, 3.9, 4.3, 4.7, 5.1, 5.6, 6.2, 6.8, 7.5, 8.2 and 9.1.

Resistors with 1% tolerance are available in the E96 series. Resistors from the E192 series have 0.5% tolerance, but they are more expensive. For completeness' sake, there also exist E3, E6 and E48 series.

If, for example, we need a 185 Ω resistor, the nearest value from the E12 and E24 series would be 180 Ω. Note that for historical reasons, circuits often use resistors with values from the E12 series only, but with a 5% tolerance.

2.2.1 Types of Resistors

Resistors are available in various shapes and configurations.

Fixed-value resistors

As the name suggests, these resistors have fixed resistance values (Figure 2.1 left). A fixed-value resistor is a two-terminal device. They can have different shapes such as rectangular, cylindrical, chassis mounted, resistor arrays, surface mounted, etc. Fixed value resistors are shown in two different ways in electrical circuits: Either as a zigzag line (US style, Figure 2.1 middle) or as a rectangle (European style, Figure 2.1, right). In a schematic, the value of a resistor is written near or inside the symbol in shorthand using multipliers like 'k' and 'M'. The Ω symbol is usually omitted. Values below 1 kΩ are commonly written with the Ω symbol replaced by 'R' (ex. 470R). Values with a decimal like 2.2 kΩ or 5.6 MΩ may be written as 2k2 or 5M6. A value of 3.9 Ω may be written as 3R9. When only references such as R1, R2, R3, etc. are shown, the values are listed in a table (the component list) or in the text, for example R1 = 10 kΩ

Figure 2.1 A fixed-value resistor (left), its US-style symbol (middle) and the European-style symbol on the right.

Variable Resistors

Also known as potentiometers ('pots') and trimmers, the value of a variable or adjustable resistor is adjusted either by turning a shaft or a screw (Figure 2.2 left) or by sliding a knob. A variable resistor is a three-terminal device. The centre pin, shown with an arrow in electrical circuits (Figure 2.2, right), is known as the wiper and is where the variable resistance is available.

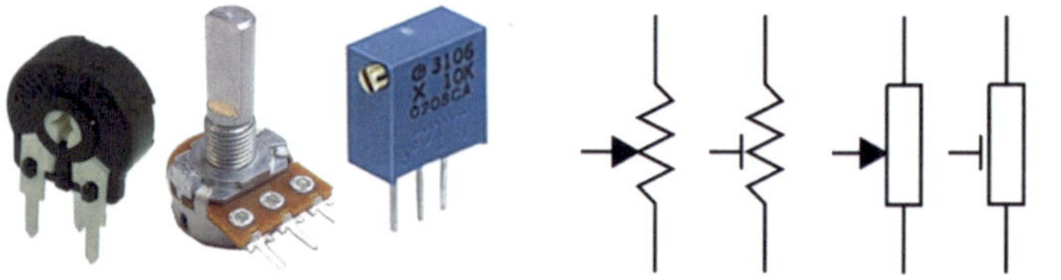

Figure 2.2 Variable resistors and their symbols. When the wiper is drawn as an arrow, it is a potentiometer. Otherwise, it is an adjustable resistor or trimmer.

Resistor Networks

Resistor networks can be used to save space when multiple resistors of the same value are needed. For such applications, it is possible to purchase resistor networks either as single inline (SIL, Figure 2.3 left) or dual inline (DIL, Figure 2.3 right) packages. They can come as multiple individual resistors, or with one pin of each resistor connected to a common contact.

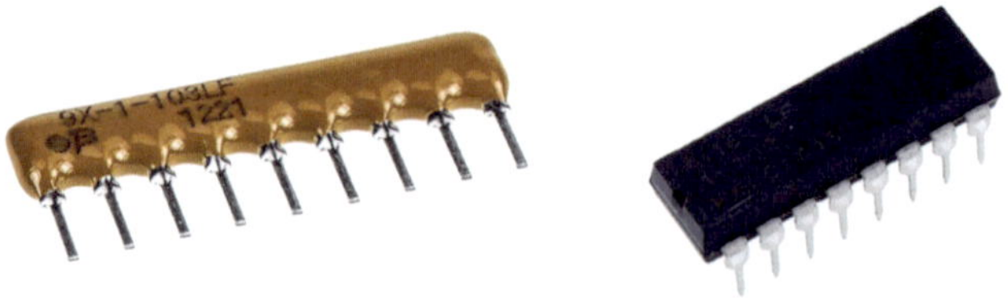

Figure 2.3 Resistor networks come in SIL (left) and DIL (right) packages.

2.2.2 Resistive Materials

Different materials are used in resistor manufacturing. Some commonly used materials, their advantages and disadvantages, are summarized in this section.

Carbon Composition Resistors

Carbon composition resistors are formed by mixing carbon granules with some kind of binding material, shaped as a small cylindrical rod. This type of resistor suffers from temperature changes, change of resistance by ageing, and high noise. Carbon composition resistors, although available in older equipment, are no longer used.

Carbon Film Resistors

These resistors (Figure 2.4a) are used for general-purpose applications. They are constructed by depositing a carbon film onto a cylindrical ceramic substrate. More carbon results in lower resistance. The advantage of this construction is low inductance, making carbon film suitable for high-frequency applications. Also, the ceramic is an excellent heat insulator.

Metal Film Resistors

Metal film resistors (Figure 2.4b) are made by layering certain metals onto an insulating substrate. Their construction is easier than carbon film resistors, and they can be manufactured in large quantities using less effort. The advantages of metal film resistors are lower noise, much tighter tolerances, and better temperature coefficients, meaning that their value doesn't change much when their temperature changes. They are slowly replacing the carbon film types.

Wire-Wound Resistors

Wire-wound resistors (Figure 2.4c) are made by winding a wire on a substrate. They are used in applications requiring very tight tolerances (less than 0.01%) and low-temperature coefficients. Wire-wound resistors are also used in high-power applications. However, due to the wound wire, this kind of resistors has high inductance and is therefore unsuitable for high-frequency applications.

Metal Strip Resistors

These resistors (Figure 2.4d) are used mainly in power supplies for measuring current. They have very low ohmic values and very low-temperature coefficients. Metal strip resistors are constructed by laser trimming a strip of metal.

Metal Oxide Resistors

Metal oxide resistors (Figure 2.4e) are similar to metal film resistors with an oxide of tin as resistive material. They are generally more suitable for high-voltage and high-power applications than metal film resistors. Metal oxide resistors are also more reliable than the metal film types.

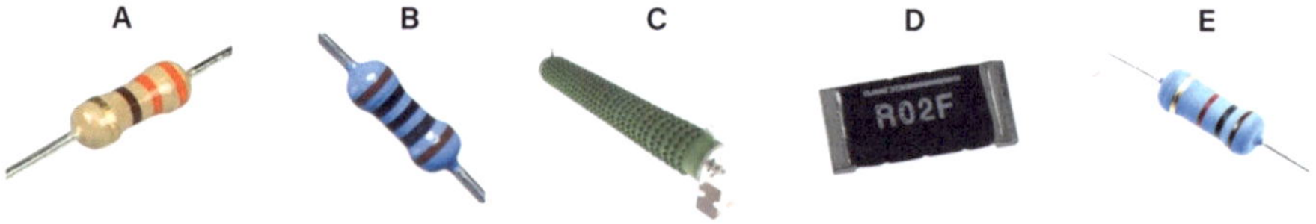

Figure 2.4 Different types of resistors. From left to right: carbon film, metal film, wire-wound, metal strip and metal oxide.

2.2.3 Resistor Specifications

It is important to know the key characteristics and specifications of resistors before they are used in electrical circuits. Some common specifications are given in this section.

Temperature Coefficient

This is a measure of the change of the resistance value due to temperature changes. The temperature can be either positive or negative. In applications where the ambient temperature is subject to large changes, resistors with low-temperature coefficients may be preferred.

Ageing

Ageing is the change of resistance value over time due to environmental stress like humidity, ambient temperature and so on. Typically, the resistance value increases over time.

Thermal Resistance

The thermal resistance is a measure of how well a resistor can dissipate heat. This parameter is important in for instance, high-power applications where the resistor is enclosed and is not in direct contact with free air.

Noise

Like all other electronic components, resistors generate noise, and this may become important in low-voltage and precision applications. Most of the noise in resistors is caused by high temperature. The noise level can be reduced by lowering the resistance value, lowering the operating temperature, and by lowering the power applied to the resistor.

Power Rating

This is an essential parameter for every resistor used in a circuit. All resistors have power ratings. In general, the larger the size of a resistor, the more power it can handle. It is important to choose a resistor that can support the maximum power applied to it. The power dissipated by a resistor is determined by the current through it and its resistance (see below). Small resistors used in most electrical circuits are designed not to dissipate more than about 0.25 W or 0.125 W.

2.2.4 Resistor Colour Code

The value of a cylindrical (or through-hole) resistor is indicated by colour bands, see Figure 2.5. Each resistor has three or four bands specifying the value, plus a tolerance band. For a 4-band resistor, the first two bands correspond to numerical values, the third band is a multiplier, and the last band is the tolerance. For a 5-band resistor, the first three bands are numerical values, the fourth band is the multiplier and the last band is the tolerance. Notice that there is a small gap between the tolerance band and other bands.

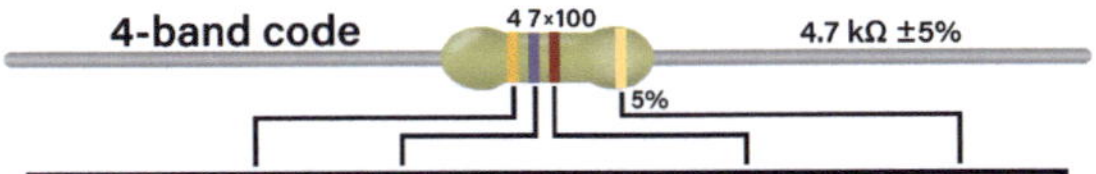

Colour	Band 1	Band 2	Band 3	Multiplier	Tolerance
None	–	–	–	–	±20
Pink	–	–	–	0.001	–
Silver	–	–	–	0.01	±10
Gold	–	–	–	0.1	±5
Black	0	0	0	1	–
Brown	1	1	1	10	±1
Red	2	2	2	100	±2
Orange	3	3	3	1,000	±0.05
Yellow	4	4	4	10,000	±0.02
Green	5	5	5	100,000	±0.5
Blue	6	6	6	1,000,000	±0.25
Violet	7	7	7	–	±0.1
Gray	8	8	8	–	±0.01
White	9	9	9	–	–

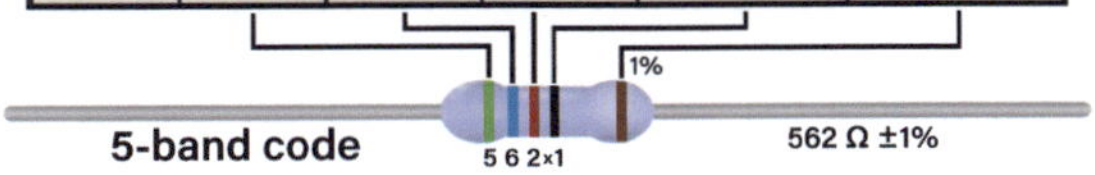

Figure 2.5 Resistor colour codes. Resistors with six bands exist too. The first five bands form a 5-band code, the 6th band indicates the temperature coefficient.

Example 1

A 4-band resistor has the colours red, red, orange, brown. Calculate its value and tolerance. From Figure 2.5 we have:

Red: 2
Red: 2
Orange: 1,000
Brown: 1%

The value is: 22 × 1,000 = 22,000 Ω or 22 kΩ with ±1% tolerance

This value can also be read from right to left, i.e. brown, orange, red, red, which corresponds to 1,300 Ω with ±2% tolerance. The reading direction is indicated by the tolerance band, which is either wider or placed apart from the other bands. In case of doubt, use an ohmmeter.

Example 2

A 4-band resistor has the colours yellow, violet, yellow, gold. Calculate its value and tolerance.

From Figure 2.5 we have:

Yellow: 4
Violet: 7
Yellow: 10,000
Gold: 5%

The value is: 47 × 10,000 = 470,000 Ω or 470 kΩ with ±5% tolerance

Example 3
A 5-band resistor has the colours: brown, black, black, brown, brown. Calculate its value and tolerance.
From Figure 2.5 we have:

Brown: 1
Black: 0
Black: 0
Brown: 10
Brown: 1%

The value is: 100 × 10 = 1,000 Ω or 1 kΩ with ±1% tolerance

This value can also be read from right to left, i.e. brown, brown, black, black, brown, which corresponds to 110 Ω with ±1% tolerance. The reading direction is indicated by the tolerance band, which is either wider or placed apart from the other bands. In case of doubt, use an ohmmeter.

Because of their small sizes, on surface-mount technology (SMT) resistors the value in ohms is printed on them as a 3- or 4-digit number. In a 3-digit representation, the first two digits are the numeric values, and the last digit is the number of zeroes. For example, 102 (Figure 2.6, left) represents '10 00' which is 1,000 Ω or 1 kΩ. Similarly, 103 represents '10 000' which is 10,000 Ω or 10 kΩ.

For low values, the letter 'R' is used as a decimal point. For example, 4R7 means 4.7 Ω (Figure 2.6, right).

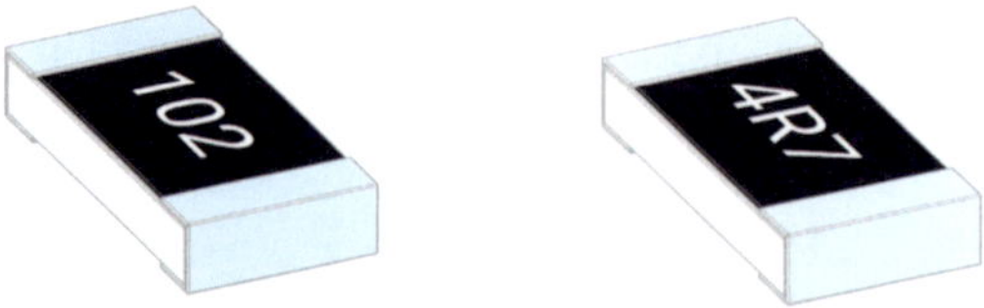

Figure 2.6 1 kΩ and 4.7 Ω surface mount resistors.

For very small-sized resistors the EIA96 system is used, which involves a lookup table where a 3-digit value is used.

2.2.5 Ohm's Law

Ohm's law is one of the fundamental laws in electronics. This law states that the voltage drop (V, in volts) across a resistor is equal to the product of the current (I, in amperes) through the resistor and its value (R, in ohms). Expressed as an equation:

$$V = I \times R \quad \text{or} \quad I = V/R \quad \text{or} \quad R = V/I$$

In many electronic circuits, the currents are measured in milliamperes (mA), which is one thousandth of an ampere. As an example, consider a circuit where the voltage across a resistor is 5 V and the current through this resistor is measured to be 100 mA. The value of the resistor can be calculated as:

$$R = V/I = 5/0.10 = 50\ \Omega$$

2.2.6 Resistors in Series and Parallel

Resistors can be connected in series, parallel, or in a combination of series and parallel. By connecting resistors together, we can obtain larger or smaller values. Also, various special passive electrical circuits can be designed by combining resistors.

Series Connection

Figure 2.7 top shows four resistors connected in series. In a series connection, the total resistance is equal to the sum of the individual values. Hence, in this circuit the total resistance R_T is given by:

$$R_T = R1 + R2 + R3 + R4$$

In a series resistor circuit, the same current flows through all the resistors and the sum of the voltage drops across each resistor is equal to the applied voltage to the circuit.

Example

Three resistors with the values 100 Ω, 150 Ω and 200 Ω are connected in series. Calculate the total resistance.

Solution

The total resistance is given by: $R_T = 100 + 150 + 200 = 450\ \Omega$

Parallel Connection

Figure 2.7 bottom shows four resistors connected in parallel. When resistors are connected in parallel, the *inverse* of the total resistance is equal to the sum of the *inverse values* of the individual resistances. Hence, in this circuit the total resistance R_T is given by:

$$1/R_T = 1/R1 + 1/R2 + 1/R3 + 1/R4$$

In a parallel resistor circuit, the same voltage is across each resistor and the sum of the currents in each branch is equal to the total current in the circuit.

Example

Two resistors with the values 100 Ω and 200 Ω are connected in parallel. Calculate the total resistance.

Solution

The total resistance is given by: $1/R_T = 1/100 + 1/200 = 3/200$

or, $R_T = 200/3 = 66.6\ \Omega$

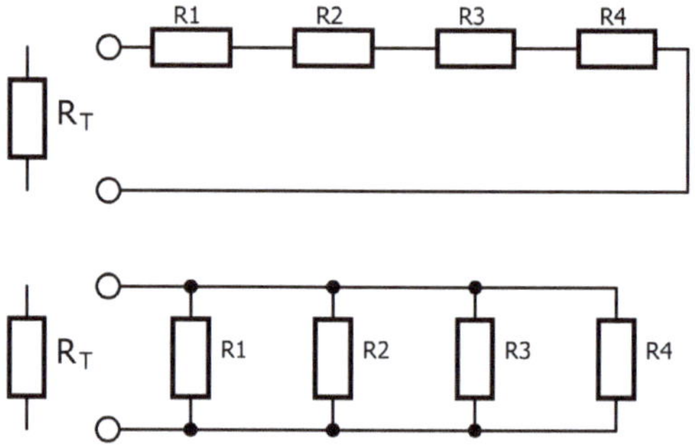

Figure 2.7 Resistors in series (top) and in parallel (bottom).

Hint: If two resistors R1 and R2 are connected in parallel, their total resistance can be calculated by multiplying the two values and then dividing by their sum:

$$R_T = R1 \times R2/(R1+R2)$$

Thus, for the above example:

$$R_T = 100 \times 200/(100 + 300) = 66.6\ \Omega$$

Hint: If two identical resistors are connected in parallel, i.e. R1 = R2, their total resistance is equal to one half of their individual resistor value, i.e. $R_T = 0.5$ R1 or $R_T = 0.5$ R2.

2.2.6 Resistors in Electrical Circuits - Examples

Some simple example circuits are given in this section to show how the total resistance, current, voltage, and power dissipation can be calculated in resistor circuits.

Example

Figure 2.8 shows an electrical circuit with a 10 V supply, and three resistors R1, R2, and R3 with values of 100 Ω, 100 Ω, and 50 Ω respectively. Calculate the current through and the voltage across each resistor. Also, calculate the power dissipation in resistor R3.

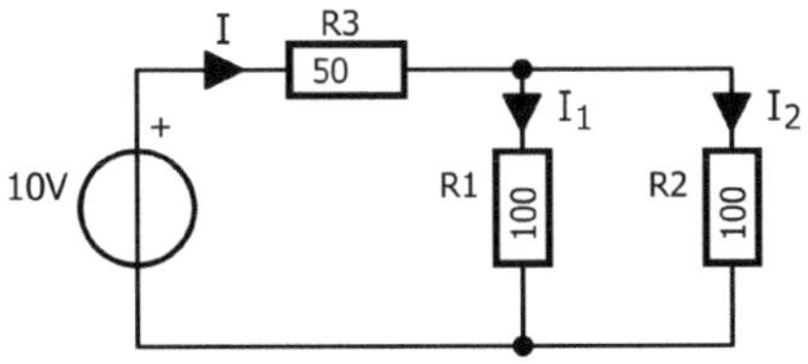

Figure 2.8 Example circuit

Solution

The sum of the currents I_1 and I_2 is equal to the current I. To find the total current I we have to find the total resistance of the circuit. R1 and R2 are in parallel, and they have the same value. Therefore, their total resistance is 50 Ω. This is in series with R3, which is also 50 Ω. The total resistance of the circuit is therefore 100 Ω. Now the current I can be calculated as:

$$I = V/R_T = 10/100 = 0.1 \text{ A or } 100 \text{ mA}$$

The voltage across R3:	$V_{R3} = I \times R3 = 0.1 \times 50 = 5$ V
The voltage across R1 and R2:	$V_{R1} = V_{R2} = 10 - V_{R3} = 10 - 5 = 5$ V
Current through R1:	$I1 = V_{R1}/R1 = 5/100 = 0.05$ A or 50 mA
Current through R2:	$I2 = V_{R2}/R2 = 5/100 = 0.05$ A or 50 mA

As you can see, the sum of the currents I1 and I2 is equal to the main current I.

The power dissipation in a resistor is given by the following formula, where P is the power in watts, V is the voltage across the resistor in volts, I is the current through the resistor in amperes, and R is the resistance in ohms:

$$P = V \times I \quad \text{or} \quad P = V^2/R \quad \text{or} \quad P = I^2 \times R$$

The power dissipated by resistor R3 is then calculated as:

$$P = V_{R3} \times I = 5 \times 0.1 = 0.5 \text{ W}$$

or $$P = V^2/R3 = 5^2/50 = 0.5 \text{ W}$$

or $$P = I^2 \times R3 = 0.1^2 \times 50 = 0.5 \text{ W}$$

Example

Figure 2.9 shows an electrical circuit with a 10 V supply and four resistors R1, R2, R3 and R4. R1 and R2 are in series, and R3 and R4 are in parallel. The two resistors in series are in series with the two resistors in parallel. Calculate the current through each resistor, the voltage across each resistor, the power dissipated by each resistor, and the total power dissipated in the circuit.

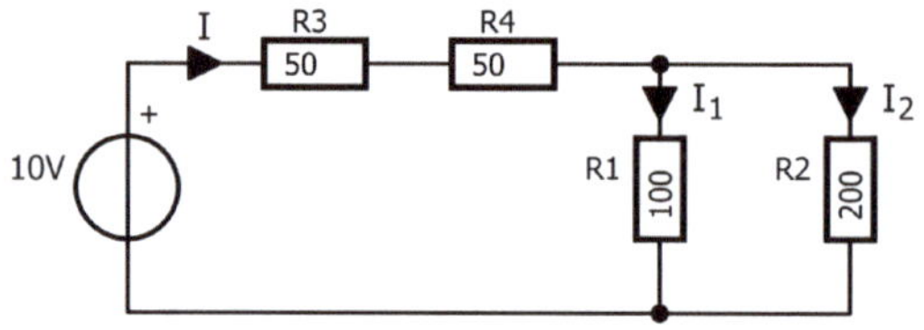

Figure 2.9 Example circuit

Solution

The sum of the currents I1 and I2 is equal to the current I. To find the total current I we have to find the total resistance of the circuit.

Total resistance of R1 and R2:	$1/R_{T12} = 1/R1 + 1/R2 = 1/100 + 1/200 = 3/200 = 66.6\ \Omega$
Total resistance of the circuit:	$R_T = 50 + 50 + 66.6 = 166.6\ \Omega$
Current I:	$I = V/R_T = 10/166.6 = 0.060$ A
Voltage over R3:	$V_{R3} = I \times R3 = 0.060 \times 50 = 3$ V
Voltage over R4:	$V_{R4} = I \times R4 = 0.060 \times 50 = 3$ V
Total voltage over R1 and R2:	$3 + 3 = 6$ V
Voltage across R1 and R2:	$V_{R1} = V_{R2} = 10 - 6 = 4$ V
Current through R1:	$I_1 = V_{R1}/R1 = 4/100 = 0.04$ A
Current through R2:	$I_2 = V_{R2}/R2 = 4/200 = 0.02$ A
Power dissipated by R1:	$P_{R1} = V_{R1} \times I_{R1} = 4 \times 0.04 = 0.16$ W
Power dissipated by R2:	$P_{R2} = V_{R2} \times I_{R2} = 4 \times 0.02 = 0.08$ W
Power dissipated by R3:	$P_{R3} = V_{R3} \times I = 3 \times 0.06 = 0.18$ W
Power dissipated by R4:	$P_{R4} = V_{R4} \times I = 3 \times 0.06 = 0.18$ W
Total dissipated power:	$P_T = 0.18 + 0.18 + 0.16 + 0.08 = 0.6$ W

Which is equal to $P_T = V \times I = 10 \times 0.06 = 0.6$ W

2.2.7 Voltage Divider Circuit

A voltage divider circuit is used to lower a voltage. For example, suppose that we want to connect a signal of 5 V to an input that cannot accept voltages above 3.3 V. We can do this by lowering the 5 V level to 3.3 V with a voltage divider.

As shown in Figure 2.10, a voltage divider circuit consists of two resistors, named R1 and R2 in the figure. If V_I is the source voltage and V_O is the required lower output voltage, we can write the following equations about this circuit.

Assuming that no or negligible current is drawn by the output (which is the case in most applications):

$$I = (V_I - V_O)/R1$$

Also, $I = V_O/R2$

Combining the two equations, we get:

$$(V_I - V_O)/R1 = V_O/R2$$

or $$R2/R1 = V_O/(V_I - V_O)$$

By selecting a value for R1, you can find the required value of R2. An example is given below.

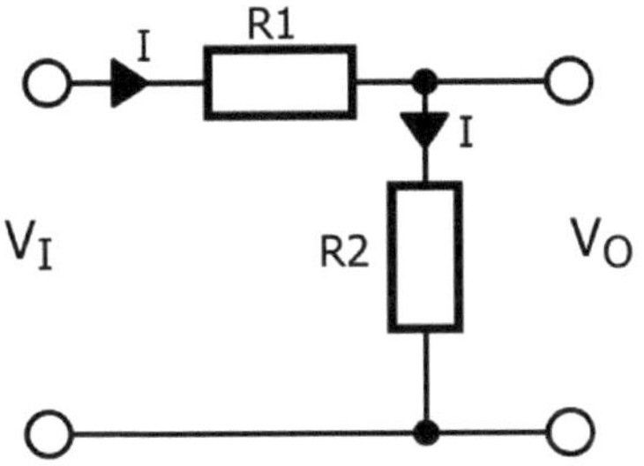

Figure 2.10 Resistive voltage divider circuit

Example
Design a voltage divider circuit to lower 5 V to 3.3 V. Assuming that R1 = 1 kΩ, calculate the value of resistor R2.

Solution
From the above equation,

$$R2/R1 = 3.3/(5 - 3.3) = 1.94$$

When R1 = 1 kΩ, then R2 = 1.94 kΩ. You can use R2 = 2 kΩ. If you don't have a 2 kΩ resistor, use two 1 kΩ resistors in series as shown in Figure 2.11. In general, we can choose any resistor with up to tens of kΩ as long as the ratio of R2 : R1 remains 2. If you want to use a value from the E12 series, you can choose between 1.8 kΩ or 2.2 kΩ. With 1.8 kΩ, V_O will be 3.2 V; with 2.2 kΩ, V_O will be 3.4 V. Therefore, to be safe, you would choose 1.8 kΩ for R2.

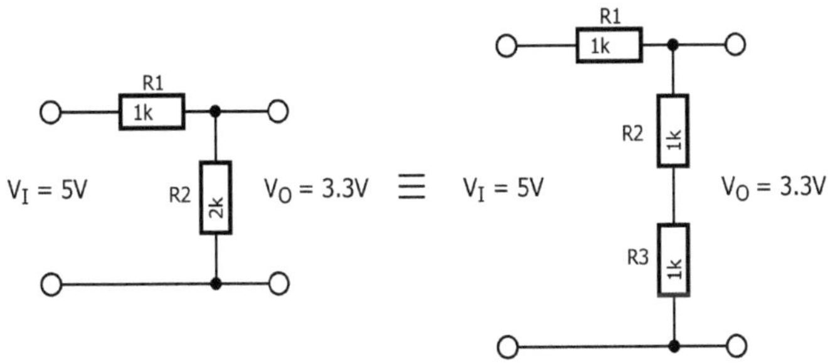

Figure 2.11 5 V to 3.3 V voltage divider.

2.2.8 Other Types of Resistors

Although the majority of resistors are standard fixed-value and variable resistors, there are a number of other resistor types that are used in specialized application. Some examples are given in this section.

Light-Dependent Resistor (LDR)

Also known as photoresistors (Figure 2.12), the resistance of an LDR changes with the intensity of the light falling on it. In complete darkness, its resistance is very large. Its resistance decreases when light falls on it. LDRs are used in many applications, such as darkness detectors and as dusk-to-dawn sensors.

Figure 2.12 Light-dependent resistor or LDR.

Thermistor

These are heat sensitive resistors (Figure 2.13) used in temperature measurement applications. Their resistance changes in accordance with the ambient temperature.

Figure 2.13 Thermistor.

Varistor

Varistors (Figure 2.14) are resistors whose resistance changes with the applied voltage. They are usually used for spike and surge protection in electrical circuits.

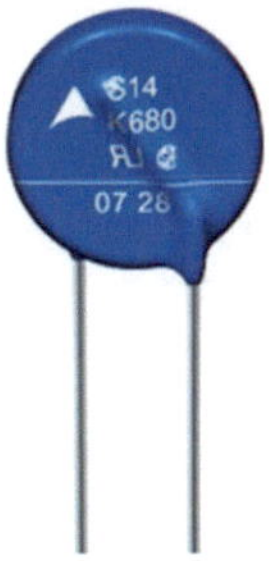

Figure 2.14 Varistor

2.2.9 Project 1 – Resistor Circuit

Aim: The aim of this project is to show how to measure current and voltage in a circuit using a multimeter.

Required components:

- 150 Ω resistor
- 120 Ω resistor
- 2 × 100 Ω resistor
- 9 V battery
- Battery cable
- Breadboard
- Jumper wires
- Multimeter

Circuit diagram: Figure 2.15 shows the circuit diagram of the project.

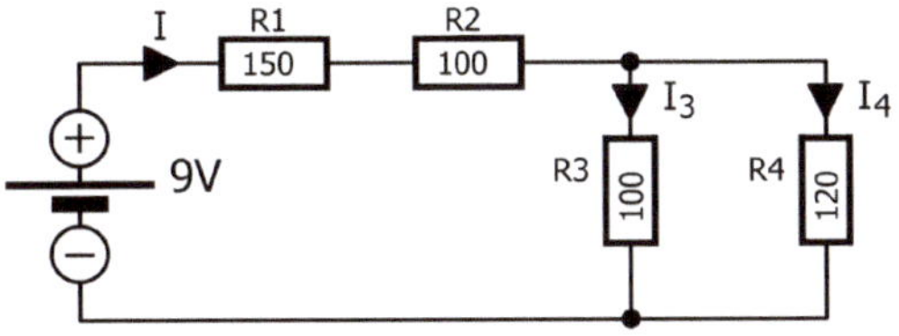

Figure 2.15 Circuit diagram of the project.

Construction: Build the project by placing the components on the breadboard as shown in Figure 2.16.

Figure 2.16 Construct the project on a breadboard.

Procedure:

- With the help of the presented theory, calculate the total resistance at the battery terminals, the current through each resistor and the voltage across each resistor.
- Remove the battery connections and measure the total resistance at the input of the circuit.
- Connect the 9 V battery as shown in Figure 2.16.
- Select voltage measurement on your multimeter and measure the voltage over each resistor by placing the voltmeter leads across the components (Figure 2.17).
- Select current measurement on your multimeter. Place the multimeter leads in series with the component whose current is to be measured (see Figure 2.17).
- Compare the theoretical values with the measured values and verify that your results are correct.

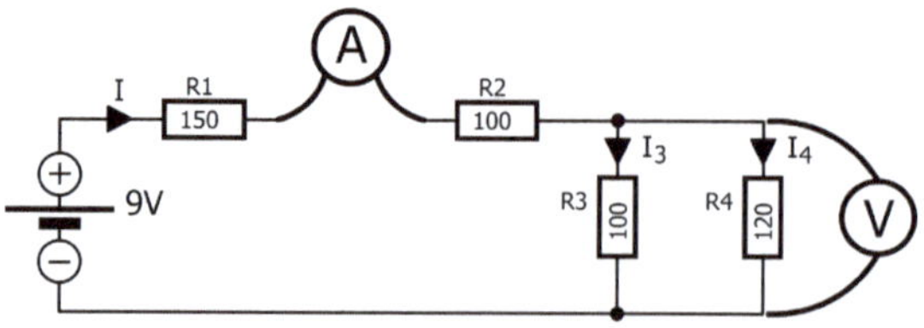

Figure 2.17 Measuring voltage and current. 'A' is a current meter, 'V' is a voltmeter.

Hint: The theoretical currents and voltages in the circuit should be as follows (there may be small differences between the theoretical values given below and the ones you have calculated because of truncation and rounding):

Total resistance R_T:	$R_T = 304.54\ \Omega$
Current I:	$I = 0.0295$ A
Voltage across R1:	$V_{R1} = 4.43$ V
Voltage across R2:	$V_{R2} = 2.95$ V
Voltage across R3 and R4:	$V_{R3} = V_{R4} = 1.61$ V
Current through R3:	$I_3 = 16.1$ mA
Current through R4:	$I_4 = 13.4$ mA

Note: Due to component tolerances, the theoretical and measured values will not be identical.

2.2.10 Using a Circuit Simulation Program

Electronic circuit simulation is a powerful tool that allows analyzing a circuit on a computer. Using a simulator program, you can draw a circuit on a computer and then display all the voltages and currents in the circuit, draw waveforms, etc. Simulation can be used for passive and active electrical circuits, as well as for digital circuits.

Circuit simulation is an invaluable tool for designing and developing electronic circuits, with many advantages:

- Any component, no matter its cost, can be modelled and simulated using a simulator. Similarly, the virtual instruments to measure signals with come for free.
- Virtual instruments and components used in a simulation cannot be damaged by incorrect wiring or by applying high voltages or currents.
- There are no calibration problems associated with virtual instruments. They are always available and operate with the same specifications.
- Simulation allows measuring internal currents and voltages that often are almost impossible to do in real life.

As powerful as simulations may be, they must be done with care. Unless you add it specifically to the simulation, effects like component tolerances, ageing, or temperature coefficients are not considered. Users may think that all components are always ideal. A simulator cannot simulate what it doesn't know about. It is only as good as what you put into it.

There are many simulation programs available. Some are free, some cost money. In this book, we will be using the freely available simulation package called TINA-TI from Texas Instruments. It can simulate all sorts of passive and active circuits as well as digital circuits, and we will use it to simulate the circuits presented in this book.

To download and install the TINA-TI simulator follow these steps:

- Go to the Texas Instruments website:
  https://www.ti.com/tool/TINA-TI

- Scroll down to TINA-TI_ENGLISH and click Download (Figure 2.18).

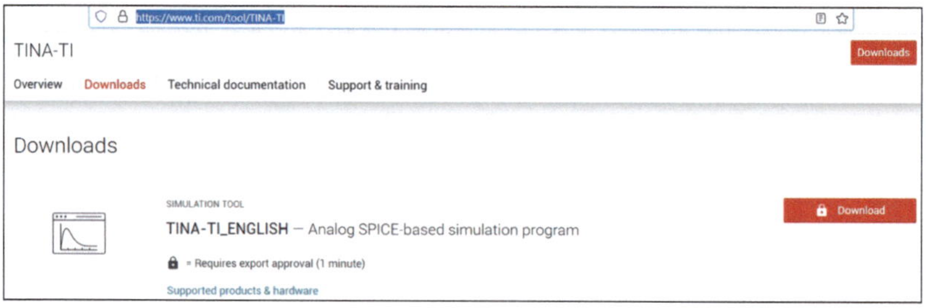

Figure 2.18 Click Download

- Log in to the site. If this is the first time you are using the Texas Instruments site, you will have to register before you can log in.

- Click to download the TINA-TI simulation package. At the time of writing this book, the package had the name TINA90-Tien.9.3.200.277.zip.

- Copy the zip file to a new folder, extract the contents, and click TINA90-Tien.exe to install the simulator program on your computer.

Simulating the Circuit From Figure 2.15

The steps are as follows:

- Click on the icon to start the simulator program. The startup screen is shown in Figure 2.19.

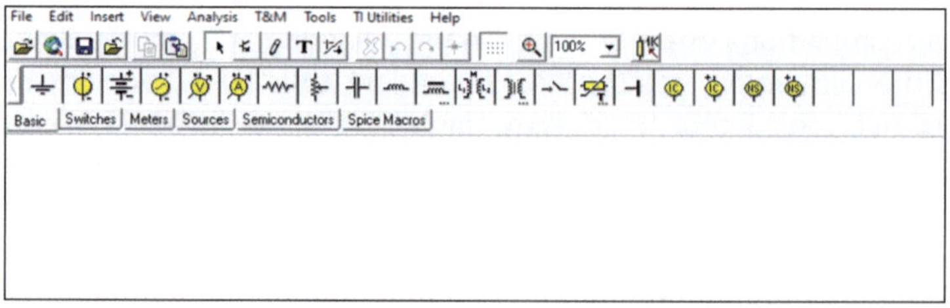

Figure 2.19 TINA-TI simulation startup screen.

- Click on resistor and place it on the screen. Click on the resistor again and join the resistors as in Figure 2.20.
- Click on battery and place the battery on the screen. Connect the battery to the circuit as shown in Figure 2.20.
- Right-click on the resistors and change their values to the values from Figure 2.15. Also, set the battery voltage to 9 V by right-clicking on it.
- Insert an earth symbol at the bottom part of the circuit. It is important that every circuit to be simulated has an earth connection.

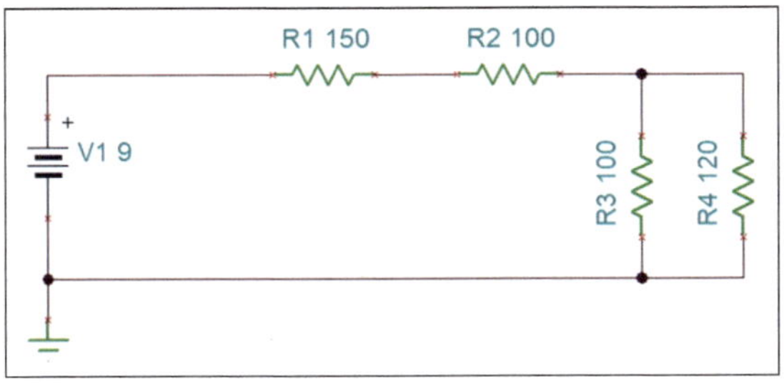

Figure 2.20 The circuit from 2.15 drawn in the simulator.

You are now ready to simulate your circuit. Click **Analysis → DC Analysis → Table of DC results**. A table of all the currents and voltages in the circuit will be displayed, as shown in Figure 2.21. **I_R1** is the current through resistor R1, and **V_R1** is the voltage across R1.

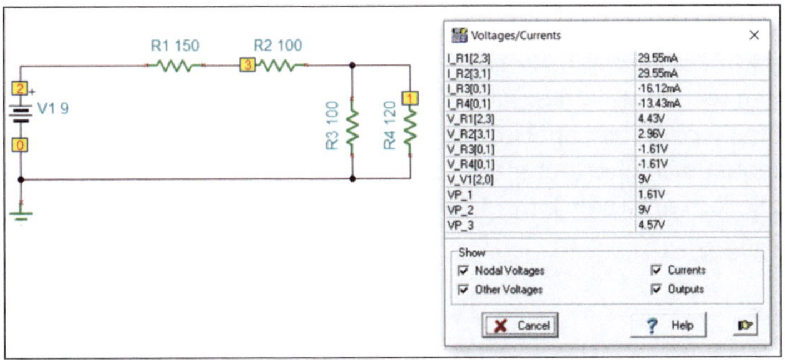

Figure 2.21 Table of the different currents and voltages in the circuit.

2.3 Capacitors

Capacitors are common components in electrical circuits. They allow AC signals to pass, but block DC current. Capacitors are passive components that store electrical energy in an electrostatic field. A capacitor is made from electrical conductors which are separated by an insulator called a dielectric. Many different dielectrics can be used in capacitors. Some common ones are teflon, polypropylene, titanium oxide, polyimide and so on.

A capacitor is described by its capacitance, which is basically the ratio of current through the device to the change of voltage over it.

$$I(t)=C\frac{dV(t)}{dt}$$

2.3.1 Types of Capacitors

There are many different types of capacitors. Some commonly used ones are described in this section.

Air Capacitor

Air capacitors are generally variable capacitors (Figure 2.22a) which use air as their dielectric. These capacitors are mainly used in resonant tuning circuits in radios, in frequency mixers, and in antenna matching circuits. The capacitance of air capacitors varies between 100 pF and 1 nF (i.e. 1,000 pF).

Ceramic Capacitor

These capacitors have fixed values, and they use ceramics as their dielectrics (Figure 2.22b). They are made with small capacitances in the range from 1 nF to 1 µF, although larger values are also available. Multi-layer ceramic capacitors (MLCC) are most often used in electronic circuits. The advantages of ceramic capacitors are their small size, and the fact that they do not have polarity, i.e. they can be connected in either direction in circuits. Ceramic capacitors are available in two classes: Class 1 capacitors are very accurate with very tight tolerances, and they are used where high stability is required. Class 2 capacitors have larger tolerances and are used in general applications.

Electrolytic Capacitor

Electrolytic capacitors (Figure 2.22c) are used where large capacitances are required. They are manufactured with tantalum or aluminium and electrolyte liquids or gels to achieve high capacitance. The electrolytic capacitor is polarised, which means that its positive terminal must be at an equal or higher potential than its negative terminal. Some disadvantages of electrolytic capacitors are their large tolerances, large leakage currents, ageing problems, and limited lifetimes.

Mica Capacitor

Mica capacitors (Figure 2.22d) are made by sandwiching mica between metal plates. Silver mica is the type commonly used. These capacitors are stable, reliable, and have low losses. They are commonly used in high-frequency circuits since their values do not change much with ageing. Mica capacitors can also be used in high-voltage applications. They're available with values between 1 pF and 4,700 pF (4.7 nF).

Film Capacitor

These capacitors (Figure 2.22e) use a thin plastic film as the dielectric. They are low-cost and have high stability with low inductance. The value of a film capacitor does not change with ageing, and it is suitable for high-voltage applications. Their inconvenience is that they tend to be bulky.

Tantalum Capacitor

Similar to electrolytic capacitors and are generally polarized (Figure 2.22f), tantalum capacitors are made of tantalum metal, allowing them to remain small. Tantalum capacitors have superior frequency characteristics than electrolytic capacitors. These capacitors are reliable, provided they are operated within their specified characteristics. It is recommended to operate them at a maximum of 50% of their rated voltages.

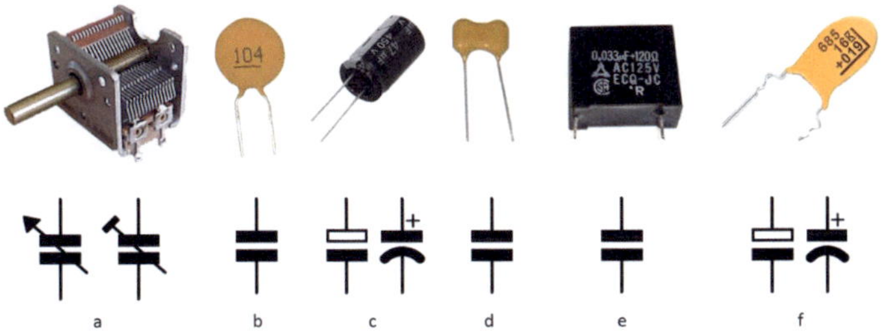

Figure 2.22 Capacitor types and their schematic symbols.

2.3.2 Capacitor Values

Capacitors are indicated with the letter 'C' in electronic circuits. The unit of capacitance is the farad (F). But, because a farad is a rather large quantity, smaller units are used in most circuits:

Unit	**Symbol**	**Farad**
Millifarad	mF	10^{-3} F
Microfarad	µF	10^{-6} F
Nanofarad	nF	10^{-9} F
Picofarad	pF	10^{-12} F

2.3.3 Capacitor Value Marking

The marking of capacitor values depends on the type of the capacitor.

Electrolytic Capacitors

As shown in Figure 2.23, the capacitance, the maximum working voltage, and the polarity are marked on the body of the capacitor. The capacitor shown has a value of 1,000 µF (1 mF) and its maximum working voltage is 16 V.

Figure 2.23 A 1,000 µF, 16 V electrolytic capacitor.

Ceramic Capacitors

These capacitors have 3-digit markings on them. The first two digits are the capacitance value, while the third digit is the multiplier (exactly as for SMT resistors), as follows:

- **0**: multiply by **1**
- **1**: multiply by **10**
- **2**: multiply by **100**
- **3**: multiply by **1000**
- **4**: multiply by **10,000**
- **5**: multiply by **100,000**
- **6**: multiply by **1,000,000**

For example, a capacitor with marking '100' is 10 × 1 = 10 pF (Figure 2.24). Similarly, '471' means 47 × 10 = 470 pF, '222' is 22 × 100 = 22,000 pF (22 nF), '104' is 10 × 10,000 = 100,000 pF (100 nF), and so on.

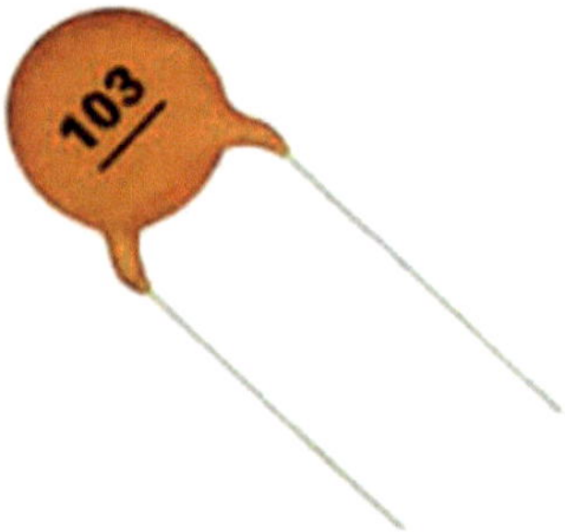

Figure 2.24 A 10,000 pF or 10 nF ceramic capacitor.

Sometimes a fourth character is printed on the capacitor as a letter and this specifies the tolerance as follows:

B: ±0.1 pF
C: ±0.25 pF
D: ±0.5 pF
F: ±1%
G: ±2%
H: ±3%
J: ±5%
K: ±10%
M: ±20%
Z: ±80%

2.3.4 Capacitors in Series and Parallel

Series Connection

Figure 2.25a shows four capacitors connected in series. In a series connection, the *inverse* of the total capacitance is equal to the sum of the *inverse* values of the individual capacitances (like resistors in *parallel*). Hence, in this circuit, the total capacitance C_T is given by:

$$1/C_T = 1/C1 + 1/C2 + 1/C3 + 1/C4$$

Example

Two capacitors with the values 100 pF and 400 pF are connected in series. Calculate the total capacitance.

Solution

The total capacitance is given by: $1/C_T = 1/100 + 1/400 = 5/400$

or $C_T = 400/5 = 80$ pF

Hint: If two capacitors are connected in series, their total capacitance can be calculated by multiplying the two capacitances and then dividing by their sum. Thus, for the above example:

$$C_T = 100 \times 400/(100 + 400) = 80 \text{ pF}$$

Hint: When two identical capacitors C1 and C2 are connected in series, their total capacitance is equal to half of one of the capacitances, i.e. $C_T = 0.5$ C1 or $C_T = 0.5$ C2.

Parallel Connection

Figure 2.25b shows four capacitors connected in parallel. In a parallel connection, the total capacitance is equal to the sum of all capacitors in the circuit (like resistors connected in *series*).

$$C_T = C1 + C2 + C3 + C4$$

Example

Two capacitors with the values 100 pF and 150 pF are connected in parallel. Calculate the total capacitance.

Solution

The total capacitance is given by: $C_T = 100 + 150 = 250$ pF

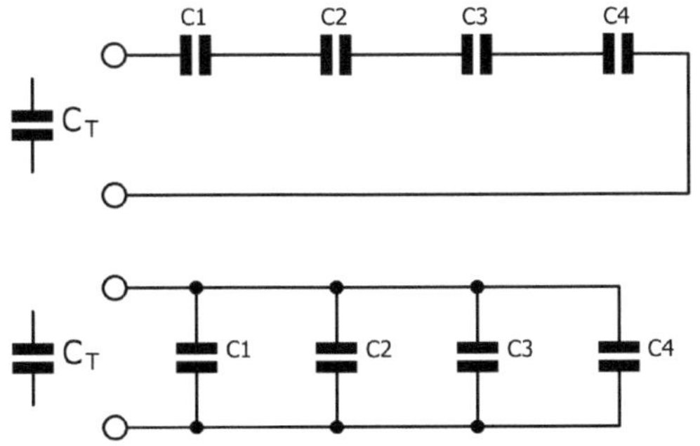

Figure 2.25 Capacitors in series (top) and in parallel (bottom).

2.4 Inductors

An inductor is a passive component in the shape of a coil. Inductors are used in electronic circuits to store energy in their magnetic fields. The most common uses of inductors are in passive electronic filters, oscillators, radio and transmitter circuits, antenna circuits, impedance matching circuits and so on. Inductors let DC current pass but have resistance (impedance) for AC signals.

An inductor is described by its inductance, which is basically the ratio of the voltage over the device to the rate of change of the current through it.

$$V(t)=L\frac{dI(t)}{dt}$$

An inductor is a coil wound on a frame. The inductance depends on the following factors:

- The shape of the coil
- Number of turns
- The spacing between the turns
- Permeability of the material where the coil is wound to (e.g. air, ferrite, iron, etc.)
- The size of the core

2.4.1 Types of Inductors

There are several types of inductors, depending on the type of material used as the core and the shape of the winding. The most common types are described briefly in this section.

Air Core Inductors

These types of inductors (Figure 2.26a) are used when the required inductance is low. An advantage of air core inductors is that there is no core loss.

Iron Core Inductors

An iron core is used in this type of inductors (Figure 2.26b). These inductors have high inductances but suffer from high-frequency effects. They are mainly used in audio applications.

Ferrite Core Inductors

In these inductors (Figure 2.26c), ferrite materials are used as the core. The type of ferrite used can be soft or hard. These inductors are commonly used in radio applications.

Chokes

A choke (Figure 2.26d) consists of a coil wound on a magnetic core. They are used in AC circuits to block high-frequency components of a waveform. Chokes are cheap since precision is not required in their manufacturing.

Toroidal Inductors

A length of wire wrapped around a donut-shaped (toroid) ferrite core (Figure 2.26e) is commonly known as a toroid core inductor. Because of its closed shape, the magnetic field is contained around the core. Toroidal inductors are used in switching power supplies, medical devices, filters, and industrial controllers.

Shielded Variable Inductors

A moving ferromagnetic core is used in these types of inductors (Figure 2.26f). The inductance is changed by moving the core in and out of a coil. These inductors are used in radio applications as intermediate-frequency-stage couplers. Typically, the inductance can be varied between 0.05 mH and 3 mH.

Chip Inductors

These consist of thin multilayer plates (Figure 2.26g) made of ferrite. The coil is printed on it with a special paste. Chip inductors are used in many electronic circuits, such as Bluetooth, Wi-Fi, wearable electronics, and so on.

Axial Lead Inductors

These inductors have a shape similar to through-hole resistors (Figure 2.26h). They feature very low DC resistance and can therefore handle high currents.

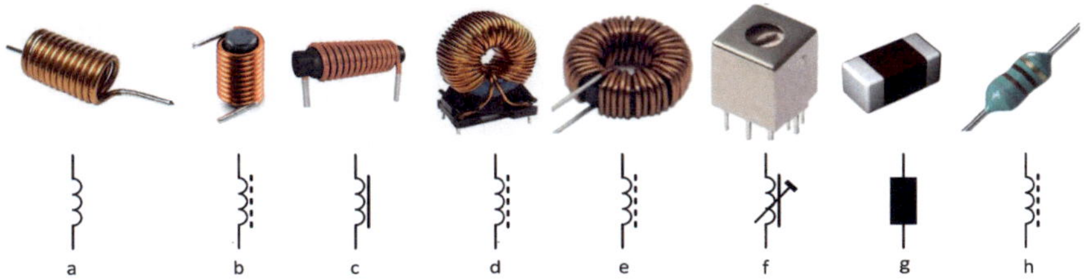

Figure 2.26 Different types of inductors and their schematic symbols. Note how the iron core is represented by a solid line, whereas the ferrite core is represented by a dashed line. The symbol below the chip inductor (h) is not typical for chip inductors and can also be used for other types of inductors.

2.4.2 Inductor Values

Inductors are indicated by the letter 'L' in electronic circuits. The unit of inductance is the henry (H). Because one henry is a large unit, smaller units are commonly used:

Unit	**symbol**	**Henry**
Millihenry	mH	10^{-3} H
Microhenry	µH	10^{-6} H
Nanohenry	nH	10^{-9} H
Picohenry	pH	10^{-12} H

2.4.3 Inductor Markings

Often, the inductance value is printed clearly on most types of inductors. If it is not printed, it can be found as described below.

Axial Lead Inductors

These inductors use the same colour code marking as resistors, and the values are found using the same colour table. The values are in microhenries (μH). An example is given below.

Example

The colours on an axial lead inductor are brown, black, red, silver. What is the inductance?

Solution

From the resistor colour code table, the inductance value can be determined:

Brown: 1
Black: 0
Red: 100
Silver: ±10% tolerance
Value = 1,000 μH ±10% (= 1 mH)

Chip Inductors

Chip inductors have a 3-digit value printed on them, followed by a letter. The first two digits are the inductance value, while the third digit is the multiplier (exactly as for SMT resistors and ceramic capacitors). The letter specifies the tolerance as follows:

F: ±1%
G: ±2%
J: ±5%
K: ±10%
M: ±20%

The letter 'R' indicates the decimal point. Examples are given below.

Example

The markings on a chip inductor are 104J. Calculate the value of its inductance.

Solution

The value is: 10×10^4 = 100,000 μH or 100 mH, ±5%

Example

The marking on a chip inductor is R12J. Calculate the value of its inductance.

Solution

'R' is the decimal point, therefore the value is: 0.12 μH, ±5%

Example

The markings on a chip inductor are 1R2K. Calculate the value of its inductance.

Solution
The value is: 1.2 μH ±10%

2.4.4 Inductors in Series and Parallel

Inductors can be connected in series, parallel, or in a combination of series and parallel. By connecting inductors together, we can obtain larger or smaller values. Also, various special passive electrical circuits can be designed by the combination of inductors.

Series Connection
Figure 2.27 top shows four inductors connected in series. In a series connection, the total inductance is equal to the sum of the individual inductances (like resistors). Hence, in this circuit the total inductance L_T is given by:

$$L_T = L1 + L2 + L3 + L4$$

Example
Three inductors with the values 100 mH, 150 mH and 200 mH are connected in series. Calculate the total inductance.

Solution
The total inductance is given by: $L_T = 100 + 150 + 200 = 450$ mH

Parallel Connection
Figure 2.27 bottom shows four inductors connected in parallel. In a parallel connection, the *inverse* of the total inductance is equal to the sum of the *inverse* values of the individual inductances (again, like resistors). Hence, in this circuit the total inductance L_T is given by:

$$1/L_T = 1/L1 + 1/L2 + 1/L3 + 1/L4$$

Example
Two inductors with the values 100 mH and 200 mH are connected in parallel. Calculate the total inductance.

Solution
The total inductance is given by: $1/L_T = 1/100 + 1/200 = 3/200$

or $L_T = 200/3 = 66.6$ mH

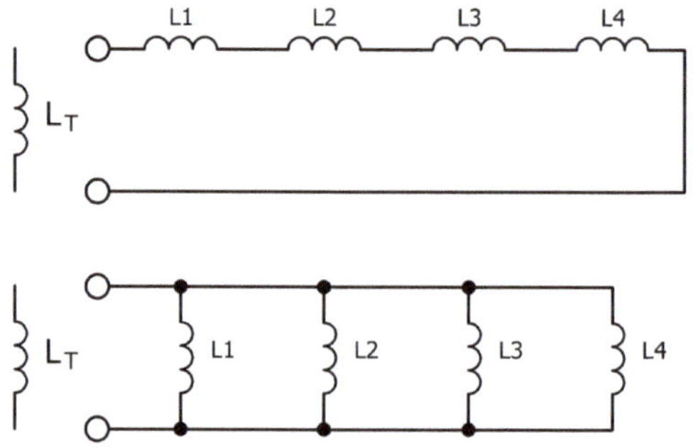

Figure 2.27 Inductors in series (top) and parallel (bottom).

Hint: If two inductors are connected in parallel, their total inductance can be calculated by multiplying the two inductances and then dividing to their sum. Thus, for the above example:

R_T = 100 × 200/(100 + 200) = 66.6 mH

Hint: If two identical inductors L1 and L2 are connected in parallel, then the total inductance is equal to half of one of the inductances, i.e. L_T = 0.5 L1 or L_T = 0.5 L2.

2.4.5 Mutual Inductance

When two coils are placed close to each other, the magnetic field produced by one of the coils generates a current in and produces a voltage over the other coil. This property is called mutual inductance. Mutual inductance is denoted by the letter 'M', and it is an important topic in radio engineering, transformers, and in the design and operation of motors. The mutual inductance between two coils is represented mathematically by:

$$M = k\sqrt{L_1 L_2}$$

Where:

M is the mutual inductance in henries
k is the coupling coefficient
L_1, L_2 are the inductances of the coils in henries

The coupling coefficient depends on how closely the two coils are coupled. It is 1 if the two coils are perfectly coupled; it is 0 when there is no coupling.

2.5 A Simple RC Network

The resistor, capacitor and inductor are often used together to create more complex circuits like adapters and filters. These circuits are often part of even more complex circuits.

Figure 2.28 shows a basic circuit consisting of a resistor, a capacitor, a DC voltage source (e.g. a battery) and a switch, all connected in series. Such so-called RC networks are extremely common in electronics circuits. What interests us here is the response of the circuit to an event, for example, the voltage across the capacitor when the switch is closed.

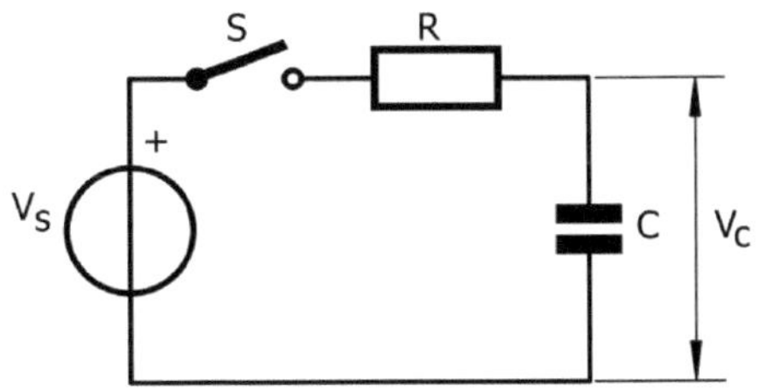

Figure 2.28 A simple and common RC network.

When switch S is open, nothing happens and the voltage over capacitor C is zero. When the switch is closed, the voltage on C starts to increase as charge (electrons) start flowing into it. However, the charging of C is not instantaneous, but the voltage rises exponentially. This is expressed by the formula:

$$V_C = V_S\,(1 - e^{-t/RC})$$

Where

V_C is the voltage across the capacitor
V_S is the battery voltage
t is the time in seconds
R is the resistor value in ohms
C is the capacitor value in farads

At time $t = 0$ (i.e. when the switch is closed), the voltage across the capacitor is 0 V. When the t is very large, the exponential term tends to zero and the voltage across the capacitor becomes equal to the applied voltage V_S.

The product of RC is known as the time constant of the circuit, and it is the time when the voltage across the capacitor rises to 63% of its final value. Thus, for example, for a 9 V supply voltage, the voltage across the capacitor reaches to 5.67 V after a time T where T = RC. Figure 2.29 shows the exponential evolution of the voltage over capacitor C.

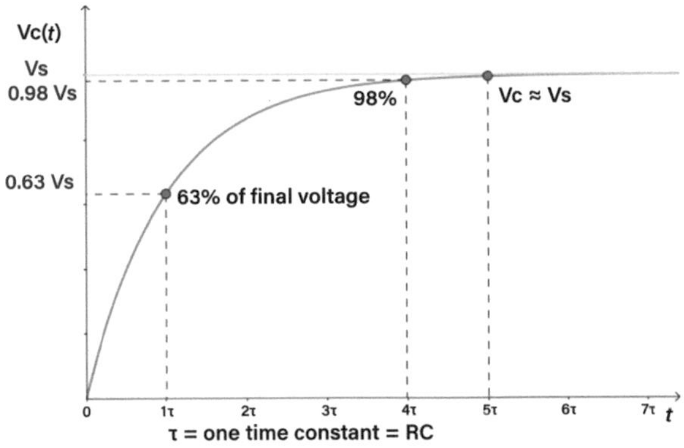

Figure 2.29 Exponential rise of the capacitor voltage.

Simulation

Let us now simulate this RC circuit using the TINA-TI circuit simulator:

- Start the simulator and draw the RC circuit shown in Figure 2.30.
- Select R = 100 kΩ and C = 10 μF by right-clicking on the components.
- Click **Switches** and insert a switch (SW-SPST1).
- Click **Meters** and select the symbol at the left and connect it to the output of the circuit. This symbol indicates the output of the circuit.

Now we are ready to simulate the circuit and draw the output waveform:

- Click the centre of the switch to close it.
- Click **Analysis → Transient**.
- Set **Start display** to 0s and **End display** to 5s. Also, click to enable Zero initial values.
- Click **OK** to draw the output waveform. This is shown in Figure 2.31.

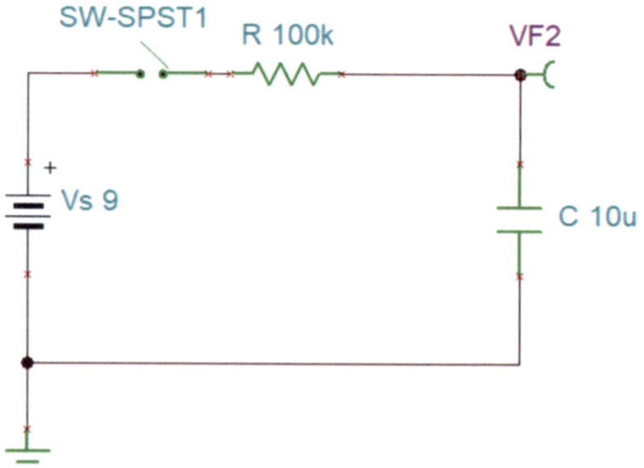

Figure 2.30 The circuit drawn in the simulator.

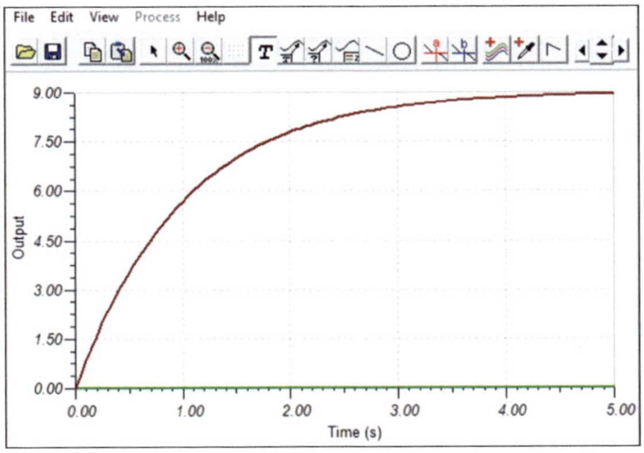

Figure 2.31 Capacitor voltage over time.

The time constant of the circuit is T = R × C = 100 × 10^3 × 10×10^{-6} = 1 second. Looking at Figure 2.31 we see that the capacitor voltage reaches 63% (5.67 V) of its final value after about one second.

2.5.1 Project 2 – Resistor-Capacitor Transient Response

Aim: The aim of this project is to show an RC transient circuit in action in a simple circuit. In this project, an RC circuit is formed, and an **active** buzzer is connected across the capacitor (an active buzzer sounds at a fixed frequency when a DC voltage is applied to it). The time constant RC is set to be 22 seconds. When the switch is pressed, the voltage across the buzzer will be about 5.67 V after 22 seconds. Therefore, the buzzer will be delayed to sound (it will sound much earlier than the time constant when enough voltage is across the capacitor to activate the buzzer). Note that once the buzzer sounds, keeping the button pressed will keep sounding intermittently since the capacitor will be partly discharged through the buzzer and then recharged.

Required components:

- 22 kΩ resistor
- 1000 µF, 25 V capacitor
- Push-button switch
- 9 V battery
- Battery cable
- Breadboard
- Jumper wires
- **Active** buzzer

Circuit diagram: Figure 2.32 shows the circuit diagram of the project.

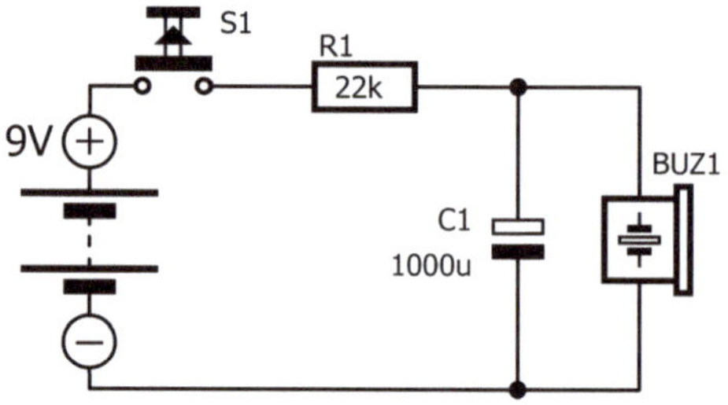

Figure 2.32 Circuit diagram of project 2.

The time constant of the circuit is given by: $T = R1 \times C1 = 22 \times 10^3 \times 1000 \times 10^6 = 22$ seconds

Construction: Construct the project by placing the components on the breadboard as shown in Figure 2.33

Figure 2.33 Assemble the project on a breadboard.

Procedure:

- Construct the circuit as shown in Figure 2.32 and Figure 2.33.
- Connect the 9 V battery.
- Press the push-button switch and keep it pressed.
- You should hear the buzzer sounding after about five seconds when the across the capacitor is high enough to activate the buzzer. Note that the buzzer will sound much earlier than the time constant of the circuit.
- Keep pressing the switch. The buzzer should sound intermittently as the capacitor is partly discharged through the buzzer and then recharged.

2.6 Resistor-Capacitor-Inductor Circuit Transient Response

As shown in Figure 2.34, an RCL transient circuit consists of a resistor, a capacitor, an inductor, a DC voltage source (e.g. battery) and a switch. Again, we are interested in analysing the response of the circuit, for example the voltage across the capacitor when the switch is closed.

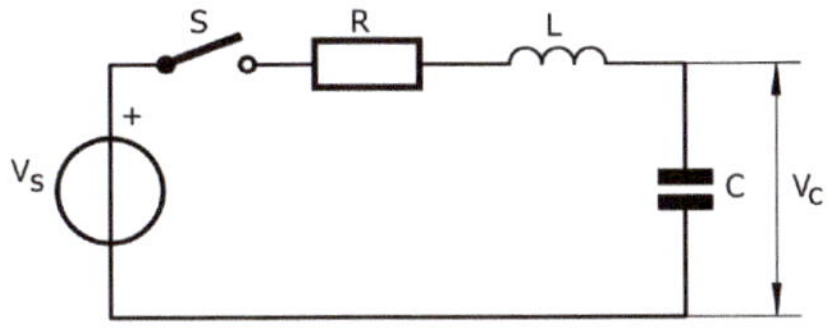

Figure 2.34 RLC circuit

When the switch is closed, the circuit can have one of three types of response depending on the component values. The important response parameter of a series RLC circuit is its damping factor, given by:

$$\zeta = \frac{R}{2}\sqrt{\frac{C}{L}}$$

Three types of responses are possible, depending on the damping factor:

Overdamped Response

When the voltage over the capacitor reaches its final value slowly without any oscillations, the circuit is said to be overdamped. This happens when the damping factor is greater than 1 (see Figure 2.35).

Critically Damped Response

The critically damped response represents the response where the output reaches its final state in the fastest possible time without oscillations. This happens when the damping factor is equal to 1 (see Figure 2.35).

Underdamped Response

When the damping factor is less than 1, the output response will show decaying oscillations (see Figure 2.35).

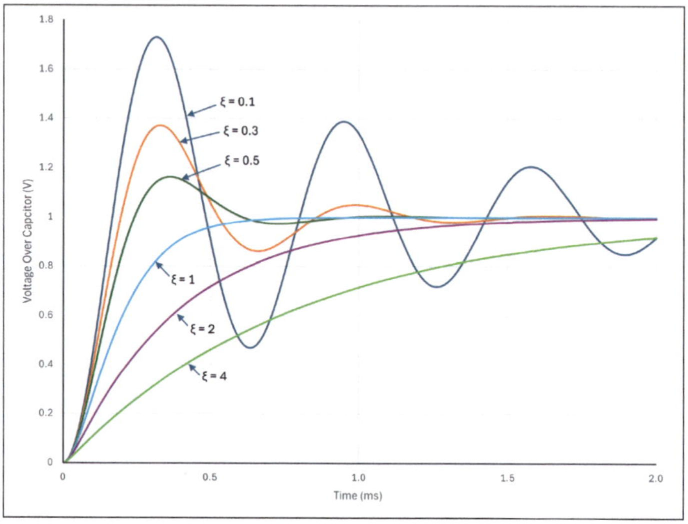

Figure 2.35 RLC transient responses.

Simulation

Let us now simulate an underdamped RLC circuit using the TINA-TI simulator. We will choose the components for a damping factor of 0.5. This gives the component values as:

R = 1 kΩ
L = 100 mH
C = 100 nF

- Start the simulator and draw the RLC circuit shown in Figure 2.36.
- Select R = 1 kΩ, C = 100 nF, L = 100 mH by right-clicking on the components.
- Click **Switches** and insert a switch (SW-SPST1).
- Click **Meters** and select the symbol at the left and connect it to the output of the circuit. This symbol indicates the output of the circuit.

Now we are ready to simulate the circuit and draw the output waveform.

- Click in the centre of the switch to close it.
- Click **Analysis → Transient**.
- Set S**tart display** to 0s and **End display** to 3m. Also, click to enable Zero initial values.
- Click **OK** to draw the output waveform. This is shown in Figure 2.37.

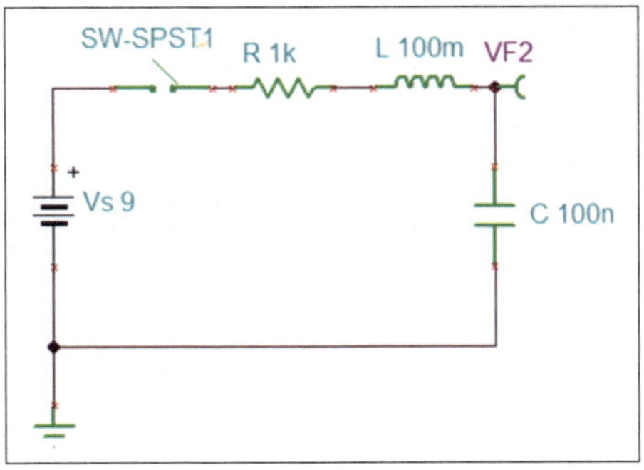

Figure 2.36 Circuit in the simulator.

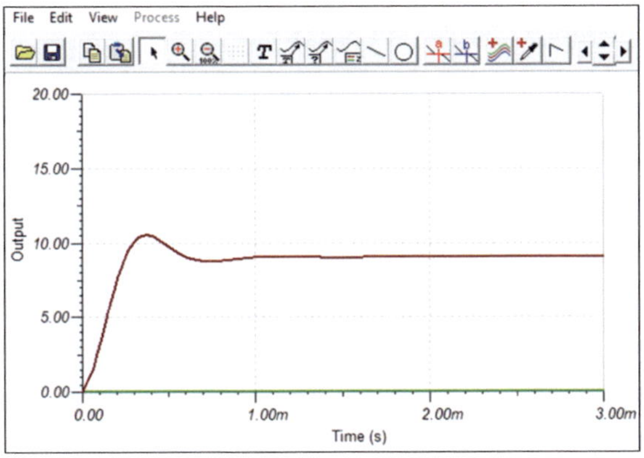

Figure 2.37 Voltage over the capacitor.

2.7 Resistance Versus Impedance

Resistors have resistance, capacitors and inductors have reactance. Both can be gathered under the common term **impedance**. Resistance and impedance are both terms related to the opposition that a circuit offers to the flow of electric current, but they have some key differences, especially in how they behave in AC (alternating current) circuits.

Impedance can be thought of as the generalization of resistance to AC circuits, taking into account the effects of capacitors and inductors. Ohm's law (V = I × R) can be extended to include impedance as follows:

$$V = I \times Z$$

Here, Z is the symbol for impedance. For an ideal resistor, Z is simply equal to R, and the value is constant no matter what the frequency of the current that flows through it is. Its impedance is 100% resistive.

For capacitors and inductors things are a bit more complicated. Their impedance is reactive, meaning that their impedance depends on the frequency of the current flowing through them. Without introducing complex calculus, the impedance of a capacitor is given by:

$$Z_C = \frac{1}{2\pi f C}$$

Here, Z_C is in ohms, *f* is the frequency in hertz and C is in farads. Similarly, the impedance of an inductor is given by:

$$Z_L = 2\pi f L$$

Here, Z_L is in ohms, the frequency *f* is in hertz and L is in henries.

Example

An ideal 100 Ω resistor has an impedance of 100 Ω at DC (0 Hz), an impedance of 100 Ω at 100 Hz, and an impedance of 100 Ω at 100 kHz.

An ideal 100 µF capacitor has infinite impedance at DC (0 Hz), an impedance of 160 Ω at 100 Hz, and an impedance of 16 mΩ at 100 kHz.

An ideal 100 µH inductor has no impedance at DC (0 Hz), an impedance of 63 mΩ at 100 Hz, and an impedance of 63 Ω at 100 kHz.

As these examples show, impedance varies with frequency in AC circuits due to the reactive components (inductors and capacitors). It is a complex quantity, often represented in the form $Z = R + jX$, where R is resistance, j is the imaginary unit ($j \times j = -1$), and X is reactance.

In the real world, **every** component, even a piece of wire, has parasitic resistance, capacitance, and inductance, resulting in an impedance, and therefore in some sort of frequency-dependent behaviour. The higher the frequency, the more pronounced the effect will be.

Chapter 3 • Magnetics

3.1 Overview

We are all familiar with magnets, also called permanent magnets. Perhaps an everyday example is a refrigerator magnet used to hold notes and items on its door. We may also use a magnet to keep our pins and needles together.

A magnet has two poles: north and south. We all know that different poles attract, and similar poles repel one another. A magnetic field is invisible, but it is often shown as lines going from the north pole to the south pole. The magnetic force is the strongest close to the poles.

Electricity and magnetics are very closely linked. Magnets are used to create electricity, and electricity is used to create magnets. Some materials, called ferromagnetic materials, can easily be magnetized. Examples of such materials are iron, nickel, and cobalt. Ferromagnetic materials can be soft like iron, and hard materials like alnico and ferrite. Both types of material can be magnetized, but hard materials can hold their magnetism whereas soft materials do not.

In this chapter, we will look at the use of magnetism in electronics.

3.2 Magnets in Electronics

Magnets are used in electronics for many applications. Electronic devices like mechanical relays, some types of microphones, loudspeakers, and transformers use magnets or magnetics. Strong magnets are used in electric motors and generators, pumps, MRI scanners, to lift heavy objects, and so on.

3.3 Electromagnets

When current flows through a wire, it generates a magnetic field around it. If we wind a wire around a metal core (e.g. iron) and then apply a voltage between the wire ends, a magnetic field is generated in the core (Figure3.1). This effect is called electromagnetism, and it is used to create electromagnets. The strength of such an electromagnet depends on the number of turns of the wire, the amount of current passing through the wire, and the properties of the core used. There are four ways to increase the strength of an electromagnet:

- More turns of wire.
- Pass more current through the wire. i.e. apply a higher voltage.
- Use a soft iron core.
- Use a bigger core.

The advantages of electromagnets are:

- Extremely high lifting forces are available.
- They can be connected in series or in parallel.
- They can be operated remotely.

Their main disadvantage is that they require a (strong) power supply.

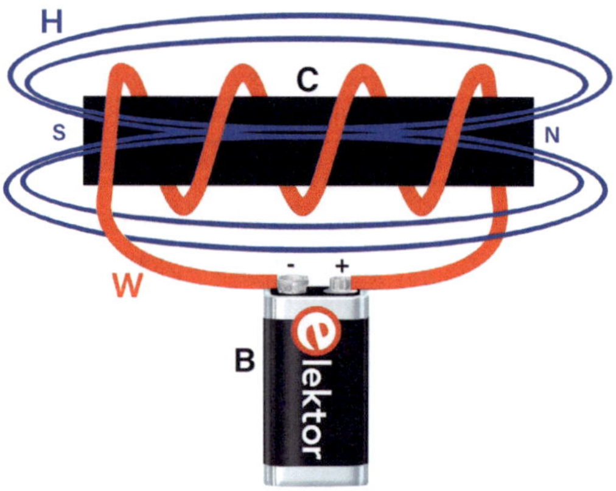

Figure 3.1 A simple electromagnet. Insulated wire W is wound around core C. When battery B is connected, the magnetic field H appears.

3.4 The Reed Switch

A reed switch is a mechanical switch with two contacts encapsulated in a glass container. The switch contacts are operated by an external magnetic field, e.g. by placing a magnet close to the contacts. Most reed switches have contacts that are normally open, but switches with normally-closed contacts exist too.

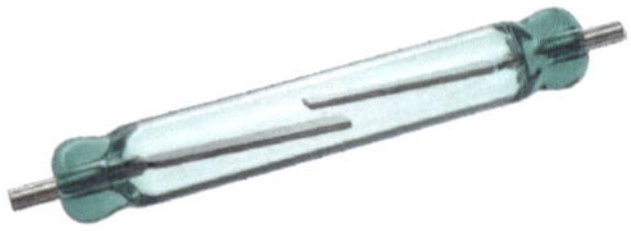

Figure 3.2 The reed switch.

Reed switches are commonly used in:

- Alarm systems to detect whether a door or a window is open or closed. A magnet is placed in or on the door or window in such a way that it is close to the reed switch when the door or window is closed. As a result, the switch contacts are closed. When the door or window opens, the magnet moves away from the reed switch and its contacts open. This is then detected by the alarm system.

- To turn a light on when, for example, the fridge door is opened.

- To prevent a device from powering on when a door or lid is not closed, for example in washing machines.

The advantages of reed switches are:

- Very low-power consumption.
- Easy to test the switch using a magnet.
- Reliable and durable.
- Low-cost.
  Very low mechanical wear and therefore a long operation life.
  It can be used in hazardous environments.
- Easy to use.

Disadvantages of reed switches:

- Can be affected by spurious electromagnetic forces.
- Vulnerable to shock and vibration.

3.5 Loudspeakers

Everyone is familiar with the loudspeaker. A loudspeaker is a device that converts electrical energy into sound energy. It is used in almost every device that generates sound, such as radio, TV, MP3 player, etc.

Most good quality loudspeakers use a coil and a magnet and are known as moving-coil loudspeakers. The device consists of a cone attached to a coil which is suspended in a magnetic field created by a permanent magnet. The audio signal generates a current in the coil and this creates a magnetic field that interacts with the static magnetic field. This causes the coil and the attached cone to move in relation to the audio signal. This movement results in audible sound (Figure 3.3).

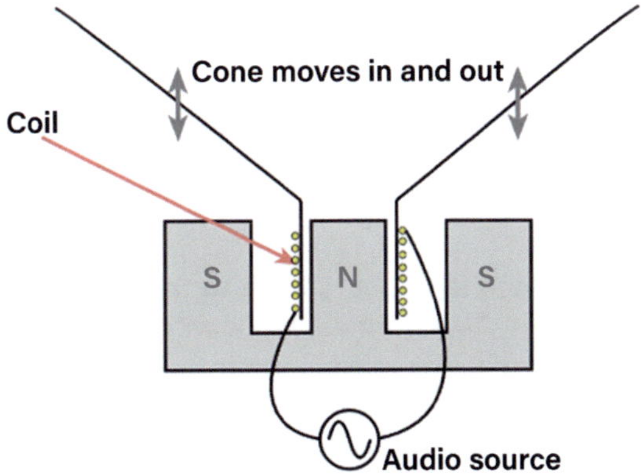

Figure 3.3 Principle of a moving-coil loudspeaker (source: www.electronics-notes.com)

Quality hi-fi equipment uses several loudspeakers, each having different frequency responses:

Subwoofer: This loudspeaker produces very-low-frequency sound in the range of 20 Hz to about 100 Hz.

Woofer: This type of loudspeaker produces low-frequency sound, usually in the range from 40 Hz to about 500 Hz.

Squawker: This loudspeaker is intended to produce mid-frequency sounds in the range of 200 Hz to about 7 kHz.

Tweeter: A tweeter produces high-frequency sound in the range of 5 kHz up to 20 kHz or higher.

In a typical hi-fi loudspeaker system, three loudspeakers are used in the same enclosure with the top one being a tweeter, the middle one a squawker, and the bottom one a woofer. Small systems with only two speakers, a squawker and a tweeter, are common too.

The impedance of a speaker is measured in ohms. Most speakers used in electronics projects have an impedance of 4 Ω or 8 Ω. The impedance is the resistance of a component in AC circuits. This is an important quantity as it determines the amount of power the speaker can handle. A 4 Ω speaker can take more energy and is therefore louder than an 8 Ω speaker. In general, low-impedance loudspeakers are used in large spaces where a high sound volume is required. High-impedance speakers are used in smaller spaces where less sound volume is required.

The power handling capacity of a loudspeaker is measured in watts. Wattage in amplifiers and loudspeakers is about how much power a loudspeaker can handle and how much power an amplifier can generate. In most electronic projects, a small 250 mW loudspeaker should be enough. Loudspeakers used in large rooms or in open-air concerts can have power capacities over hundreds or thousands of watts.

Note that loudspeakers do not like DC current. Applying a DC voltage to a loudspeaker will very likely result in a burned coil.

3.6 Transformers

A transformer is a passive component that is used to transfer electrical energy from one part of a circuit to another part. Transformers work with AC voltages, and they are used to step up or down an AC voltage. The output voltage of a step-up transformer is higher than its input voltage. A step-down transformer lowers its input voltage. For example, a step-down transformer can be used to decrease the 230 V_{AC} mains voltage to 9 V_{AC}. A transformer provides galvanic isolation between two parts of a circuit. They are also used in electronic circuits to adapt the impedance of an output to the impedance on an input.

Figure 3.4 shows the structure of a transformer. Basically, it consists of wires wound on opposite sides of a rectangular-shaped iron core. One side is called the primary, which is the input side, while the other side, the output, is called the secondary of the transformer.

The magnetic flux created by the input AC voltage at the primary generates the voltage at the secondary side of the transformer.

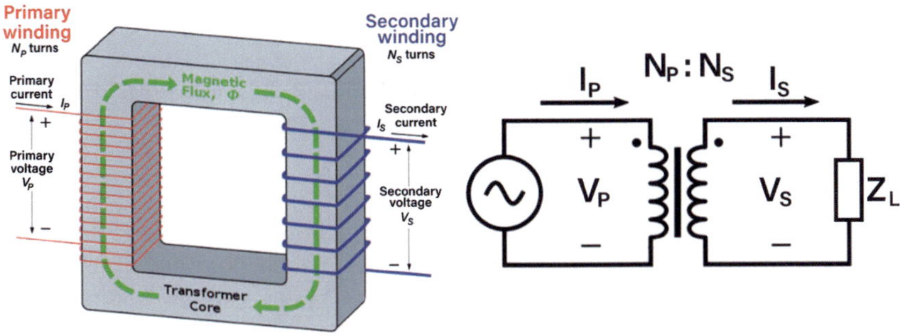

Figure 3.4 The transformer and its use in a circuit. Note how the primary and secondary coils are wound in the same way around the core. This is indicated by the dot in the schematic representation on the right (source: en.wikipedia.org).

An ideal transformer has no losses. The voltage at the output of an ideal transformer depends on the applied input voltage and the ratio of the turns at each side. If N_P and N_S are the number of turns at the primary and secondary respectively, the output voltage V_S is then given by:

$$V_S = V_P \frac{N_S}{N_P}$$

In a step-up transformer, $V_S/V_P > 1$; in a step-down transformer, $V_S/V_P < 1$.

Example

The number of primary and secondary turns in a transformer are 150 and 300 respectively. If the input voltage is 12 V, calculate the output voltage assuming there are no losses in the transformer.

Solution

From the above equation, $V_S = 12 \times 300/150 = 24$ V.

Transformer Losses

In a real transformer, there are several types of losses. Therefore, it is not 100% efficient. Some of the most important losses are:

- Hysteresis losses due to nonlinear magnetic effects in the core.
- Eddy current losses due to current induced in metal parts of the transformer.
- Losses due to winding resistance.

- Magnetic flux losses as some of the primary magnetic flux does not pass the secondary.

- Capacitive effects between windings.

The efficiency of a transformer is denoted as a percentage, and it is equal to the ratio of its output power divided by the input power, multiplied by 100.

Isolation Transformer

Isolation transformers are used to isolate primary-side circuits from secondary-side circuits physically and electrically (i.e. galvanically), usually for safety reasons. As well as providing safety, isolation transformers also eliminate or diminish high-voltage spikes in supply lines that may cause damage to equipment. Some voltage spikes can be as high as several thousand volts, and these can easily damage electrical equipment connected to them. An isolation transformer reduces these spikes considerably before they reach the load. They are also used in electrical engineering laboratories for safety reasons.

Current Sensing Transformer

The current sensing transformer is used to sense and measure currents in for instance a power supply. The electromagnetic field produced by the current is sensed with a coil and the actual current is estimated. Current transformers are used to prevent overcurrent conditions in sensitive circuits. The current clamp uses this technique to measure currents in mains wiring.

Auto Transformer

An auto transformer is an electrical transformer with only one winding. The same winding is used as the primary and secondary winding. Its operation is like a transformer with two windings, but it is different in the way the primary and secondary windings interact. Figure 3.5 shows an auto transformer with its circuit diagram.

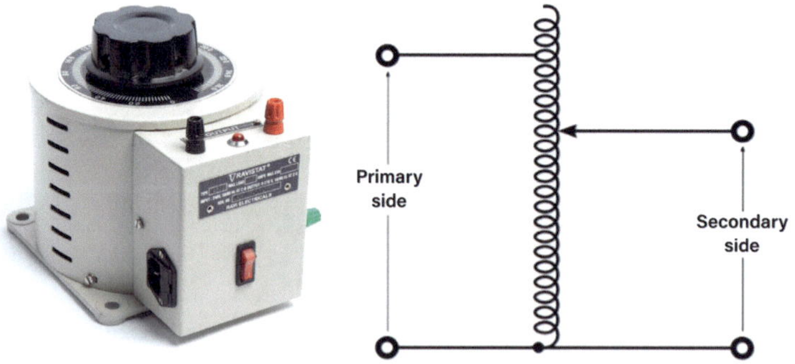

Figure 3.5 An adjustable auto transformer and its circuit.

The advantages of the auto transformer are:

- Smaller than a standard transformer, saving material. Therefore, an auto transformer can be cheaper.
- Better voltage regulation properties due to less wire being used, resulting in lower resistance.
- Better efficiency due to lower ohmic resistance and less magnetic material.

The inconveniences of the auto transformer are:

- Because the low- and high-voltage circuits are connected together, the low-voltage circuit must be designed to withstand the higher voltages.
- There is no galvanic isolation between its primary and secondary windings.
- Only suitable for small turn ratios.

3.7 Relays

A relay is like a switch, but with its contacts controlled by an electrical signal. Basically, it is an electrically operated switch. A relay lets you use a small amount of power to switch a large amount of power. Relay contacts come in the same configurations as switches. i.e. SPST, SPDT, DPST, DPDT, etc. Figure 3.6 shows the circuit symbols of different types of relays.

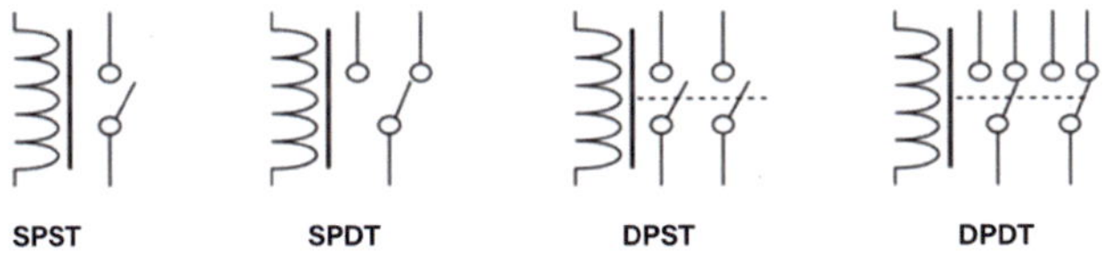

Figure 3.6 Circuit symbols for different types of relays.

Broadly speaking, there are two types of relays as far as their operation is concerned: electromechanical and solid-state.

Electromechanical Relays

Most high-power relays are electromechanical types. These relays work on the principle of electromagnetic induction. As shown in Figure 3.7, a coil is wound on a soft iron core. The contacts of the relay can either be normally closed (NC) or normally open (NO). When current flows through the coil (e.g. when voltage is applied to the coil), an electromagnetic filed develops in the core which pulls on the contacts, thus changing the state of the relay. When the current is removed, the relay contacts return to their original states by a spring action. When an electromechanical relay is used in transistor circuits, it is necessary to connect a reverse-biased diode in parallel with the relay to avoid any damage to the transistor.

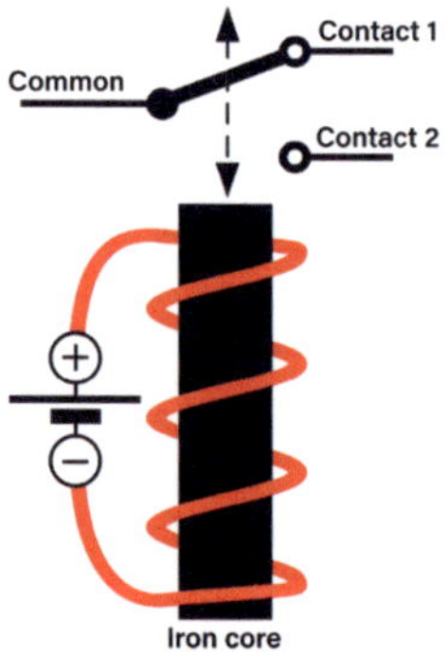

Figure 3.7 Principle of operation of an electromagnetic relay.

Figure 3.8 shows some examples of electromechanical relays.

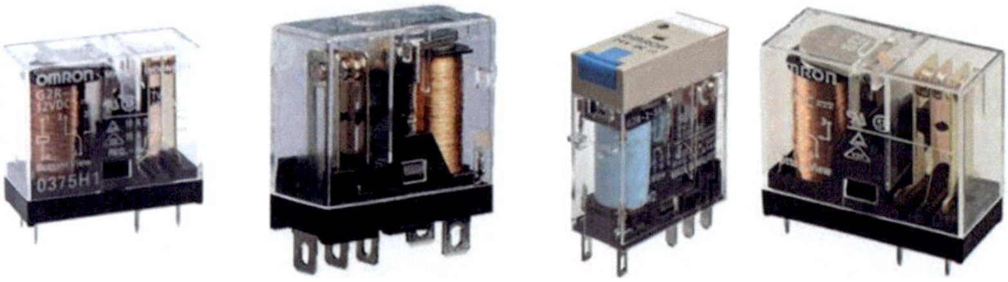

Figure 3.8 Some electromechanical relays.

Solid-State Relays

Solid-state relays (SSR) have the same purpose as electromechanical relays, but they do not have coils or moving parts. These relays are designed using semiconductor parts, and they can be thought of as being the semiconductor equivalent of electromagnetic relays. Just like a normal electromechanical relay, SSRs provide galvanic isolation between their input and output contacts. Solid-state relays can be designed to switch both AC and DC currents. In electronic circuits, currents are often small and hence solid-state relays are preferred unless a heavy load must be switched such as a heater, a fan, a motor, etc.

The main advantage of solid-state relays over conventional electromechanical relays is that they have no moving parts. There is no contact wear and there are no contact-bounce problems to worry about. Table 3.1 compares electromechanical relays and solid-state relays.

	Electromechanical Relay	Solid-State Relay
Power consumption	Usually large	Small
Operation current	Medium to large	Small
Shock and vibration	Moderately reliable	Very reliable
Electrical noise	Large switching noise	No switching noise
Switching speed	Moderate (ms)	Fast (μs)
Environment	May not operate in magnetic environments	No environmental problems
Size	Small to medium	Can be very small
Contact wear	Yes	No

Table 3.1 Electromechanical relays vs. solid-state relays

Figure 3.9 shows some solid-state relays.

Figure 3.9 Examples of solid-state relays.

Chapter 4 • Active Components

4.1 Overview

Active electronic components are devices that can introduce gain or control the flow of electric current in electronic circuits, enabling them to perform various functions essential for electronic systems and devices. The most common types of active electronic components are diodes, transistors, and integrated circuits (ICs or chips). These components can perform operations like switching and amplifying signals, regulate power, perform calculations, store data and so on.

Semiconductors are made by stacking layers of n-type and p-type semiconductor material on top of each other. The region between two different layers is called a p-n junction. N-type material contains negative charge carriers (electrons), while p-type material contains positive charge carriers (holes). The conducting properties of these layers can be controlled by applying so-called bias voltages and currents.

In this chapter, we will be looking at some of the most used active components: diodes, transistors, and some ICs.

4.2 Semiconductor Diodes

When we mention diodes in this chapter, we mean semiconductor diodes made from silicon or germanium. They are also called p-n-junction diodes, as they consist of one p-n junction. A diode is a two-terminal device that conducts current in one direction only. Its resistance is very low (ideally zero) in one direction and high (ideally infinite) in the other direction. For a diode to conduct current, its anode terminal must be positive regarding its cathode terminal. This is called forward biasing.

Some uses of diodes are:

- Demodulation in radio receivers
- Rectifiers to convert AC signals to DC
- As small light sources (e.g. light-emitting diodes)
- Constant voltage sources (e.g. Zener diodes)
- Clamping circuits
- Clipper circuits
- Logic gates
- Voltage multipliers
- Switches
- Surge protectors
- Flyback circuits
- Reverse voltage protection
- Etc.

4.2.1 Diode Types

There are several types of diodes. Some common ones are described in this section.

Signal Diode

These are small diodes used in many applications to limit the current flow in one direction only. Figure 4.1 shows a typical (small) signal diode with its symbol in an electrical circuit.

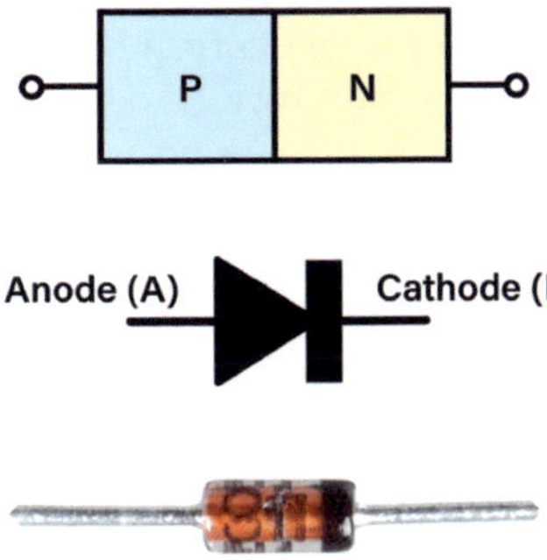

Figure 4.1 PN-junction-type small-signal diode.

Figure 4.2 shows the I-V characteristic of a signal diode. It does not conduct until its forward bias voltage V_F is about 0.7 V for silicon diodes (0.3 V for a germanium diode). Care should be taken not to exceed the maximum diode specifications, such as the maximum forward current IF and the maximum reverse voltage V_R that can be applied to a diode before breakdown occurs. Maximum power dissipation is also an important diode specification that must be respected. The total power dissipated by a diode can be determined by multiplying the voltage over it (normally 0.7 V for silicon diodes) by the current flowing through the diode.

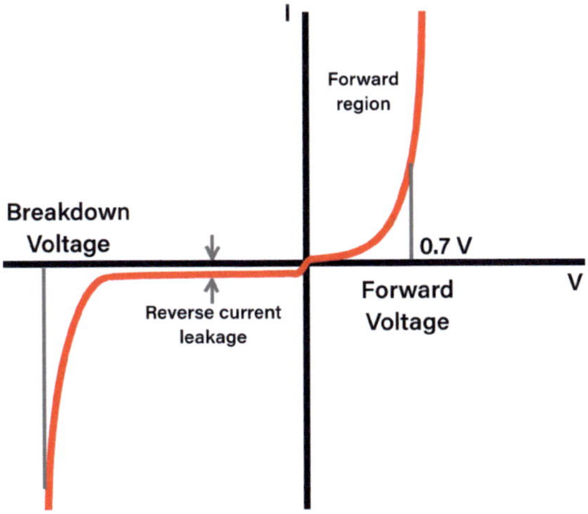

Figure 4.2 Signal diode current-voltage (I-V) curve.

As an example, the 1N4148 is a widely used signal diode in all sorts of applications. The specifications of this diode are as follows:

- Maximum reverse voltage V_{RM} = 100 V
- Average forward current I_F = 200 mA
- Maximum forward current I_{FM} = 300 mA
- Maximum forward voltage drop V_{FM} = 1.0 V at 10 mA
- Non-repetitive peak forward surge current I_{FSM} = 1.0 A
- Total power dissipation P_D = 500 mW
- Reverse recovery time t_R < 4 ns

Schottky Diode

Schottky diodes are very similar to signal diodes. The forward voltage drop of the Schottky diode is 0.2 V to 0.3 V. They are used in fast switching circuits. Figure 4.3 shows the circuit symbol of the Schottky diode. The disadvantage of Schottky diodes is that they have low reverse breakdown voltages.

Figure 4.3 Schottky diode schematic symbol.

Rectifier Diode

Rectifier diodes are similar to signal diodes, but their p-n junction area is very large. These diodes are used in AC-to-DC rectifier circuits. Rectifier diodes have the same circuit symbols and shapes as signal diodes.

Tunnel Diode

A tunnel diode, also called an Esaki diode, exhibits a region of operation with negative resistance (rdiff < 0), meaning that the current decreases as the voltage increases. When the voltage is first applied to the diode, current stars flowing through it. The current increases with the increasing voltage. Once the voltage rises high enough, the current starts to decrease. Increasing the voltage higher starts increasing the current again. Tunnel diodes are used in switching circuits, oscillators, and as logic memory devices. Figure 4.4 shows the circuit symbol of a tunnel diode and its typical I-V characteristic.

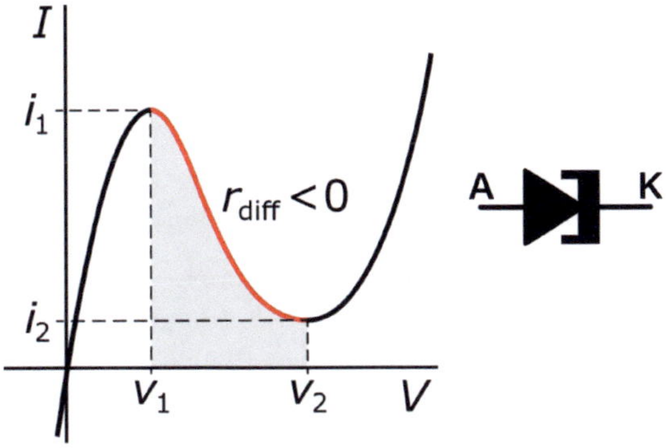

Figure 4.4 Tunnel diode.

Photodiode

A photodiode is a light-sensitive semiconductor diode that converts light to electricity. Photodiodes are used in many applications such as generating small amounts of electricity (solar cell), in smoke and fire detectors, in cameras as light sensors, in medical instruments to analyse samples, and so on. Figure 4.5 shows a photodiode and its circuit symbol.

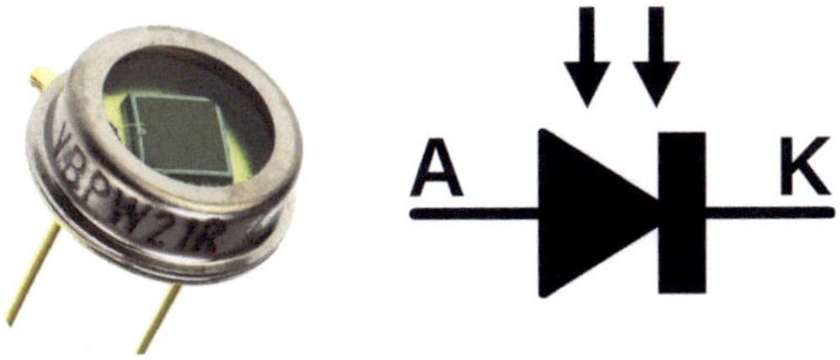

Figure 4.5 Photodiode.

Varactor Diode

This is a diode whose internal capacitance varies depends on the applied reverse voltage. It is used in circuits that require variable capacitance. For example, it is used in radio tuning circuits where the capacitance of the tuning circuit can be adjusted by a voltage. The varactor diode always works in reverse bias, and it is a voltage-dependent device. The symbol of a varactor is similar to that of a signal diode, except that a parallel plate is added to its anode terminal.

Zener Diode

Zener diodes allow current to flow in both directions when certain conditions are met. When forward biased, Zener diodes work in the same way as signal diodes. When connected in reverse-bias mode, no current can flow (except for a small leakage current). However, when the reverse voltage over the Zener diode becomes higher than the Zener breakdown voltage, current starts flowing through the diode. From this point on, the voltage over the diode remains constant over a wide range of the applied voltage, independent of the current flowing through the diode. Zener diodes are used in voltage regulation, voltage references, surge suppression, switching circuits, and in clipper circuits. Figure 4.6 shows the schematic symbol of the Zener diode and its I-V characteristic.

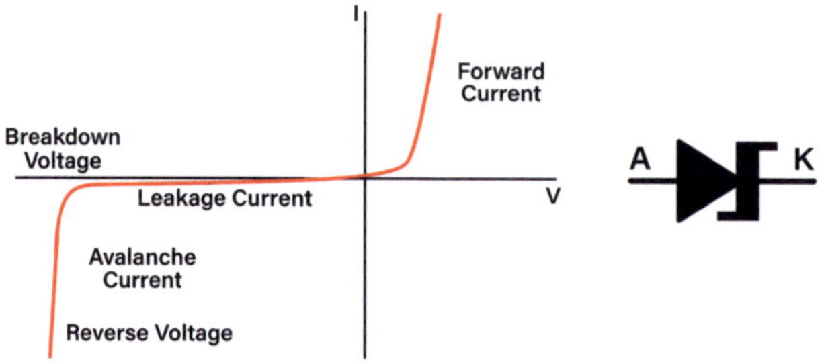

Figure 4.6 The Zener diode and its I-V characteristic.

Zener diodes are available with different standard voltages from as low as 2.4 V up to hundreds of volts.

Light-emitting diode (LED)
A light-emitting diode is a semiconductor diode that emits light when current flows through it. LEDs are used in many electronic circuits. Some common applications are:

- Flashlights & lighting
- Computers & clocks
- Traffic signals (signage)
- Indicator lamps
- Televisions
- Vehicle head lamps
- Street lighting
- Etc.

LEDs have many interesting properties, here are a few:

- Fast operation
- Different colours
- Produce little heat
- Small
- Brightness can easily be controlled
- Low cost
- No warm up period
- Can be clustered easily

LEDs are available in several colours. The brightness of an LED is controlled by the amount of current passing through it. The forward voltage V_F of an LED depends on the colour of the light it emits. As shown in Figure 4.7, the V_F voltage drop is between 1.8 V and 3.5 V.

LED colour	Forward voltage
Red	1.8 V
Yellow	2.1 V
Green	2.2 V
Blue	3.2 V
White	3.2 V

Figure 4.7 Forward voltages of LEDs.

LEDs are available in various sizes. Most LEDs are cylindrical, although rectangular and other shape LEDs such as strips are also available (Figure 4.8). Standard through-hole LEDs are 3 mm and 5 mm in diameter. No matter the size of the LED, it usually requires about 16–20 mA for full brightness (even though 1 mA is often bright enough). LEDs intended for lighting applications accept much higher currents. High-efficiency LEDs only need a few milliamperes of current.

Figure 4.8 LEDs.

RGB LED

An RGB LED (Figure 4.9) consists of three LEDs, red, green, and blue, in the same package. You can produce almost any colour by combining those three colours (Figure 4.10).

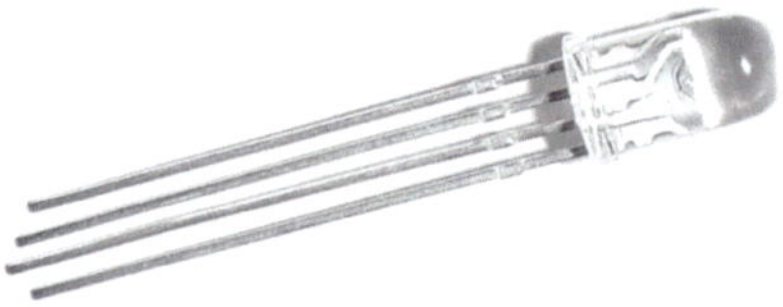

Figure 4.9 RGB LED.

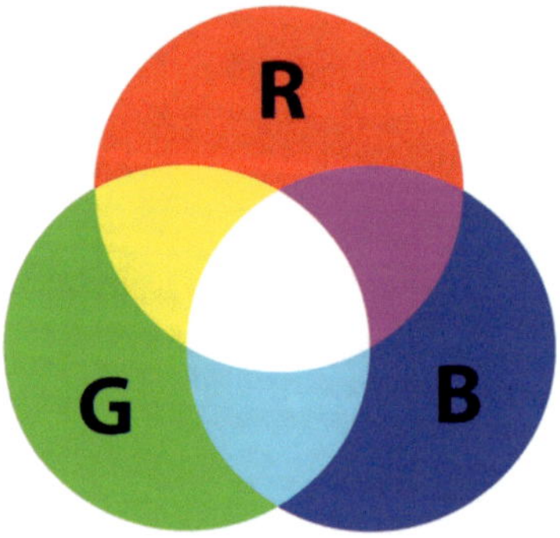

Figure 4.10 Producing different colours from RGB.

RGB LEDs are available in two types: common-cathode (CC, Figure 4.11 left) and common-anode (CA, Figure 4.11 right). The common terminal is longer than the other three terminal for identification. If it is not known, the type of the RGB LED can easily be identified using a multimeter, as follows:

- Put your multimeter in diode-tester mode.
- Connect the red multimeter lead to the longest lead (common lead) of the RGB LED.
- Connect the black lead of the multimeter to any other lead of the LED.
- If the LED lights up, it is a common-anode (CA) type. If not, it is a common-cathode (CC) LED (or it is broken).

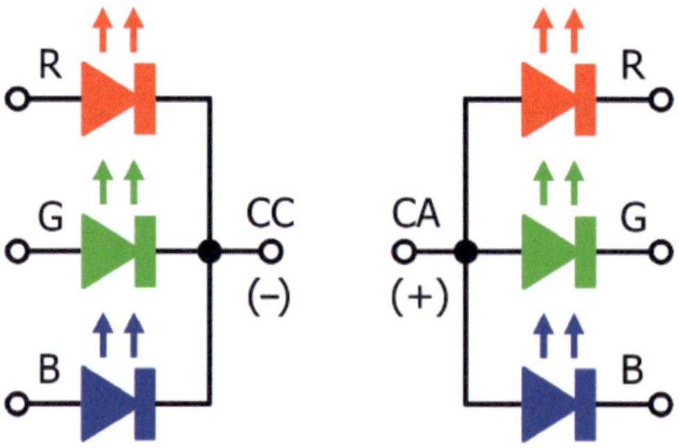

Figure 4.11 RGB LED types.

Bicolour LED

Two or bicolour LEDs (Figure 4.12) are three terminal devices, with usually one common-cathode (CC) terminal and two anode terminals for each colour (CA types exist too, of course). The common terminal is the middle terminal for identification.

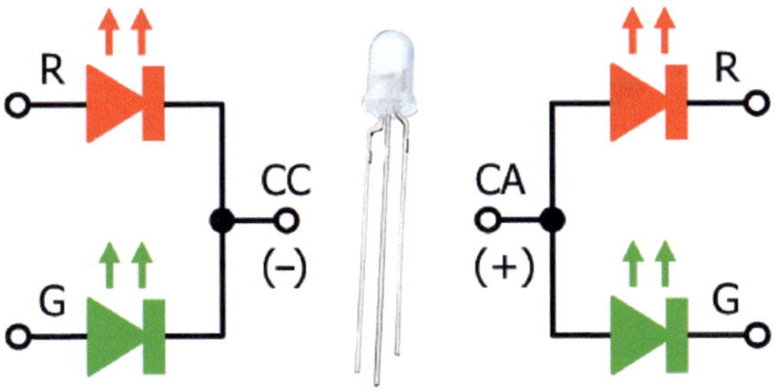

Figure 4.12 Bicolour LED.

Besides the LEDs mentioned above, there also exist infrared and ultraviolet LEDs, high-voltage LEDs (several LEDs in series in one device), digital LEDs, colour-changing and blinking LEDs, etc. Note that a red LED will always emit red light, but a water clear transparent LED may emit any colour, often red, blue, or white. Use a diode or LED tester to find out which colour.

4.2.2 Diode Circuits

In this section, various diode circuits and diode-based projects are given.

4.2.2.1 Simple LED Circuit

Figure 4.13 shows a simple LED circuit. In this circuit, a 9 V battery and a standard 5 mm red LED are used together with a current limiting resistor. In this circuit, we will pass 15 mA of current through the LED, which should be quite bright. The value of the current limiting resistor is calculated as follows:

If the red LED's forward voltage V_F (i.e. the voltage across the LED) is 1.8 V, the voltage over the resistor will be 9 – 1.8 = 7.2 V. Using Ohm's law, the value of the resistor is now calculated as:

$$R = V/I = 7.2/0.015 = 480\ \Omega$$

The nearest E12-series standard value is 470 Ω. Using this value results in a current of 7.2/470 = 0.0153 A = 15.3 mA, which is close enough (2% error).

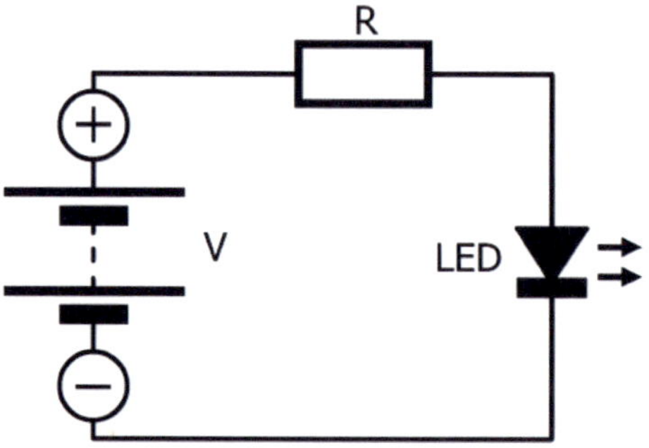

Figure 4.13 Simple LED circuit.

As 15 mA produces rather bright red light, you may want to decrease the LED current to about 3 mA. To achieve this, the value of R must be recalculated: R = 7.2/0.003 = 2,400 Ω or 2.4 kΩ. You can use the nearest E12-series of 2.2 kΩ for a slightly higher brightness, or 2.7 kΩ for a slightly lower brightness.

Note that the brightness of light as perceived by the human eye is not linear, but logarithmic. As a rule of thumb: To increase the brightness of an LED by one step, the current through the LED must be doubled.

4.2.2.2 Project 1 – LED With Push-Button

Aim: The aim of this project is to show how to control an LED with a switch. The LED turns on whenever the push-button is pressed.

Required components:

- 2.2 kΩ resistor
- Low-current red LED
- Push-button switch
- Battery cable
- Breadboard
- Jumper wires
- 9 V battery

Circuit diagram: Figure 4.14 shows the circuit diagram of the project.

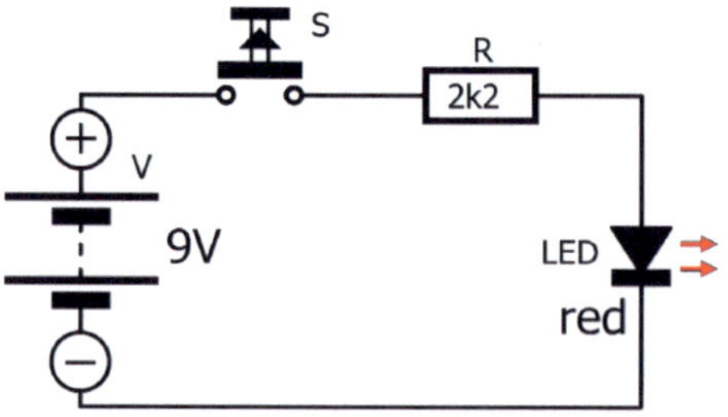

Figure 4.14 Circuit diagram of the simple LED circuit.

Construction: Construct the project by placing the components on the breadboard, as shown in Figure 4.15.

Figure 4.15 Construct the project on a breadboard.

Procedure:

- Construct the circuit and connect the battery
- Push the button. You should see the LED shine as long as the button is pressed.

4.2.2.3 Project 2 – RGB LED With Push-Buttons

Aim: The aim of this project is to show the operation of an RGB LED circuit. Three buttons are used, with each button connected to one colour of the LED. Pressing a button turns on the corresponding LED. Different colours can be obtained by pressing more than one button at the same time.

Required components:

- 3× 1 kΩ resistors
- Common-cathode RGB LED
- 3× Push-button switches
- Battery cable
- Breadboard
- Jumper wires
- 9 V battery

Circuit diagram: Figure 4.16 shows the circuit diagram of the project. 1 kΩ resistors are used in this project for moderate brightness. Smaller resistors may be used for higher brightness if required. But not too small (330 Ω minimum), or you will break the LED.

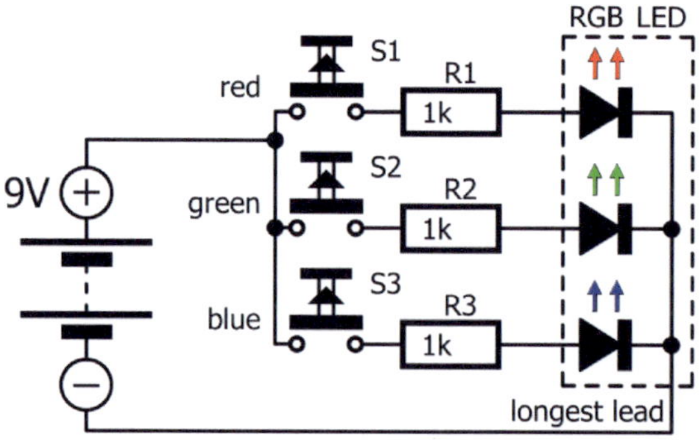

Figure 4.16 Circuit diagram of the project.

Construction: Construct the project by placing the components on the breadboard as shown in Figure 4.17.

Figure 4.17 The simple RGB LED circuit built on a breadboard.

Procedure:

- Construct the circuit by connecting the longest lead of the RGB LED to the negative terminal of the battery wire and connect the battery. Make sure that your RGB LED is a common-cathode (CC) type. If not, change the polarity of the battery connections.
- Push a button and the corresponding LED colour will turn on. Press more than one button at the same time to create different colours.

4.2.2.4 Project 3 – Zener Diode Regulator

Aim: The aim of this project is to design a Zener-diode-based voltage regulator. In this project, a 9 V battery is used as the source voltage. A 5.1 V, 1 W Zener diode (V_Z = 5.1 V) is used to drop the voltage to 5.1 V for the load. It is assumed that the load current can vary between 0 ma (i.e. no load) to the maximum that can be supported by the chosen Zener diode power rating.

Figure 4.18 shows the circuit diagram of a Zener-diode-based voltage regulator, where the output voltage is available across the Zener diode. The value of the series resistor RS will be calculated in the design steps given below.

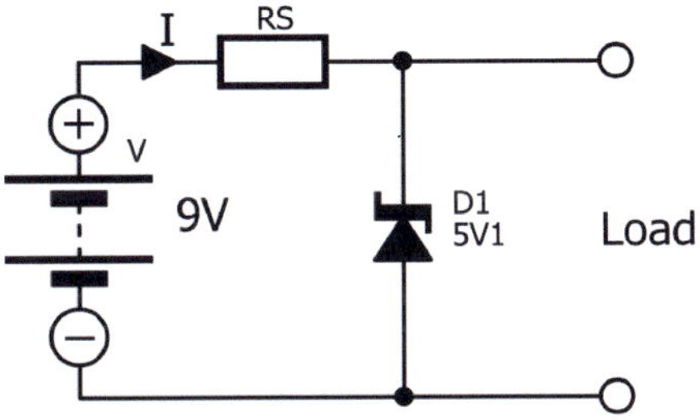

Figure 4.18 Zener-diode-based voltage regulator.

The design steps are as follows:

- The maximum current I_z through the Zener diode, allowing for 50% of power derating for diode safety: P/V = 0.5/5.1 = 0.098 A = 98 mA.

- The maximum value of the series resistor $R_S = (V - V_Z)/I_Z =$ (9 – 5.1)/0.098 = 39.8 Ω. We will use R_S = 39 Ω as the nearest value.

Normally, we allow 10% of the load current for the Zener diode. Therefore, the maximum current the load can draw is about 88 mA.

When there is no load, the maximum current through the Zener diode will be (9 – 5.1)/39 = 100 mA. The maximum power dissipated in the diode will be 100 × 5.1 = 510 mW, which is well within the specification of the Zener diode. The maximum power dissipated by the R_S is 0.098 × (9 – 5.1) = 382 mW. Allowing for twice this power for resistor safety, a 1 W resistor should be fine.

Required components:

- 39 Ω, 1 W resistor
- 5.1 V, 1 W Zener diode
- Battery cable
- Breadboard
- Jumper wires
- 9 V battery

Construction: Assemble the circuit from Figure 4.18 on a breadboard, with R_S = 39 Ω.

Procedure:

- Connect the 9 V battery to the circuit
- Measure the voltage across the Zener diode using a multimeter in voltage measurement mode to confirm that it is 5.1 V.
- Draw some load currents by placing various resistors across the Zener diode and measure the voltage to make sure that the voltage remains 5.1 V. For example, connect 100 Ω for a load current of 51 mA.

Simulation

The steps are:

- Start the TINA-TI simulator.
- Draw the circuit as shown in Figure 4.19 by selecting the MZP4733 as Zener diode model. This is a 5.1 V Zener diode.
- Connect a 100 Ω resistor at the output as the load resistor.
- Click **Analysis → DC analysis → Table of DC results**
- A table will be displayed showing all the voltages and currents in the circuit. Note that the voltage across the load (V_{F1}) is 5.1 V and the current through the load (I_R1) is 51 mA (see Figure 4.20)

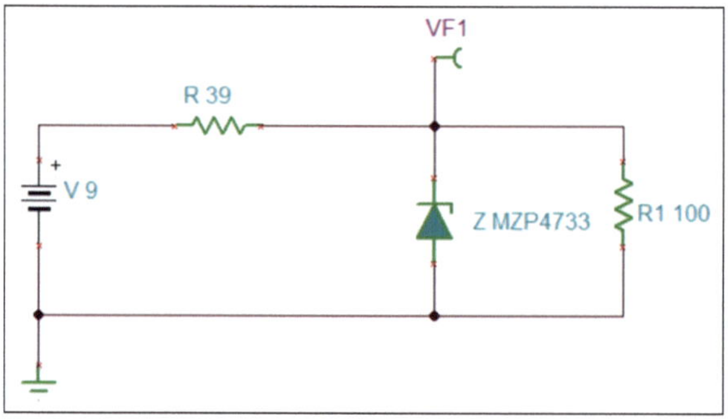

Figure 4.19 Simulation circuit.

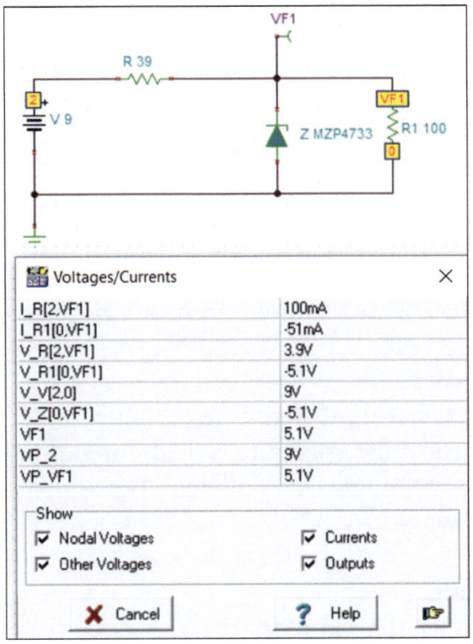

Figure 4.20 Results of the circuit simulation.

Example

In this example, we will design a Zener-diode-regulator circuit assuming that the load current is a constant 500 mA. Assume that the source voltage is 9 V and the required load voltage is 5.1 V.

Solution

Allowing 10% of the load current value as Zener current, the total current through resistor R_S in Figure 4.18 will be 550 mA. Therefore, the power dissipation in the Zener diode will be 5.1 × 0.05 = 255 mW. Allowing for a safety margin of approx. 50%, we will use a 0.5 W diode.

The required value of the series resistor is R_S = (9 – 5.1)/550 = 7.1 Ω. We can use a 6.8 Ω resistor as the nearest value. The power dissipation in the resistor is 0.55×(9 – 5.1) = 2.1 W. We will use a 5 W type for its safety.

Zener Diodes in Series

Zener diodes can be connected in series to obtain higher voltages. An example circuit is shown in Figure 4.21 consisting of 5.1 V Zener diodes.

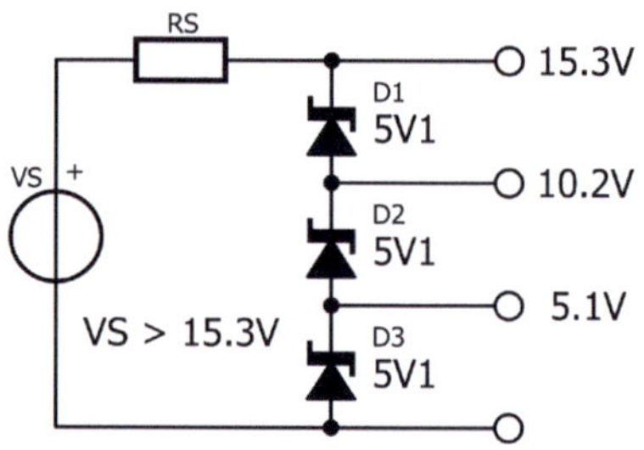

Figure 4.21 Zener diodes can be connected in series.

4.2.2.5 Rectifier Circuits

Rectifier circuits are used to convert AC signals to DC. Electronic equipment usually operates from DC voltages. Therefore, most mains-powered equipment incorporates one or more rectifier circuits. These can be made with diodes. There are two main types of rectifier circuits: half-wave and full-wave.

Half-Wave Rectifier Circuit

A half-wave rectifier circuit is the simplest rectifier circuit, consisting of a single diode. Figure 4.22 shows the circuit diagram. In most applications the mains supply is the input, a sine wave, and a step-down transformer is used to lower the voltage to the required value. The output consists of the positive part of the input sine wave because the diode only conducts in one direction. The output peak voltage is 0.7 V lower than the transformer output due to the voltage drop V_F across the diode.

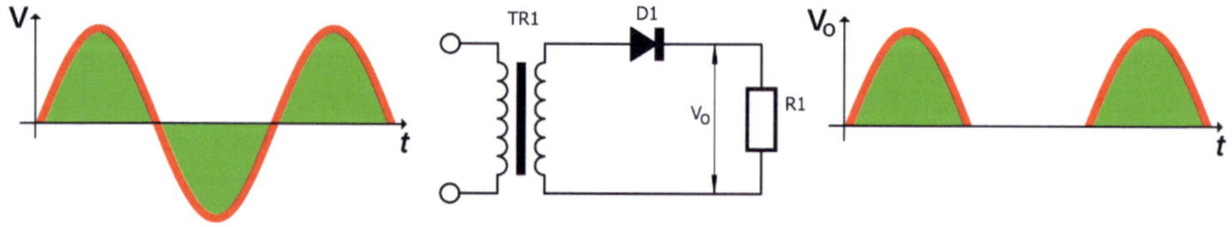

Figure 4.22 Half-wave rectifier circuit.

Sometimes a large capacitor is placed in parallel with the load. The effect of this capacitor is to smooth the large ripples as shown in Figure 4.23.

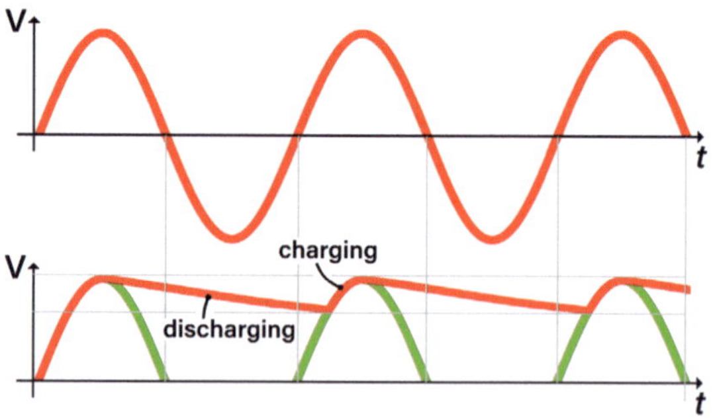

Figure 4.23 Waveforms with a smoothing capacitor placed after the diode.

Simulation

Perhaps the easiest way to understand the operation of a half-wave rectifier circuit is to simulate it and place various load capacitors across its output. The steps are as follows:

- Start the TINA-TI simulator and draw the circuit shown in Figure 4.24.
- Set a Voltage Source as the input and set it to sine wave with an amplitude of 9 V and a frequency of 50 Hz (Figure 4.25).
- Insert the monitoring points VF1 and VF2 so that the input and output waveforms can be plotted. Connect a 100 Ω resistor as the load.
- Click T&M and select oscilloscope. Set Channel to VF1, Time/Div to 10m, Volts/Div to 5. Then, set Channel to VF2 and use the same settings. Move the input graph up using the Position button. Figure 4.26 shows the input and output waveforms of the half-wave rectifier circuit displayed on the oscilloscope.

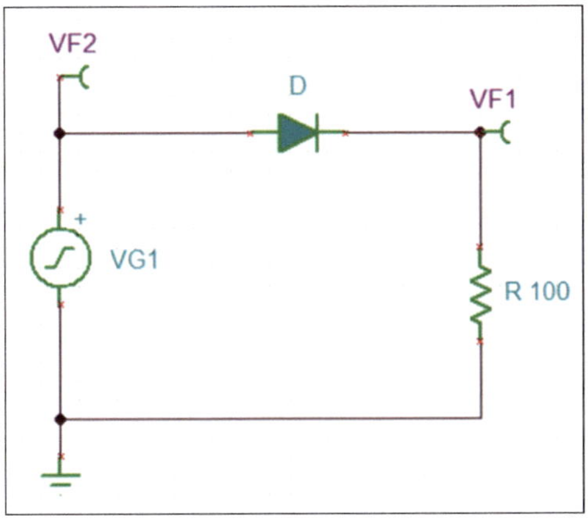

Figure 4.24 Circuit diagram.

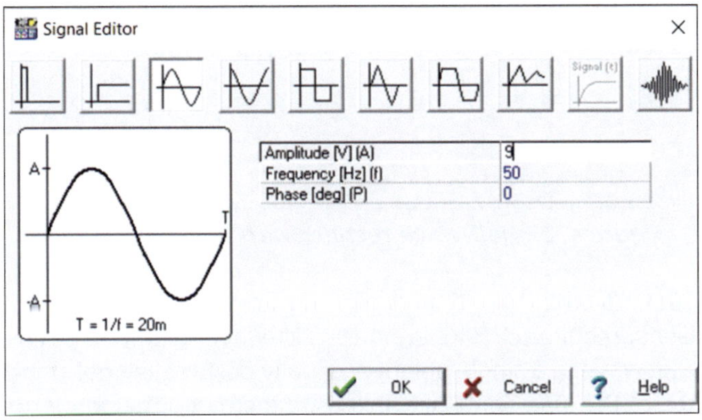

Figure 4.25 Use a Voltage Source.

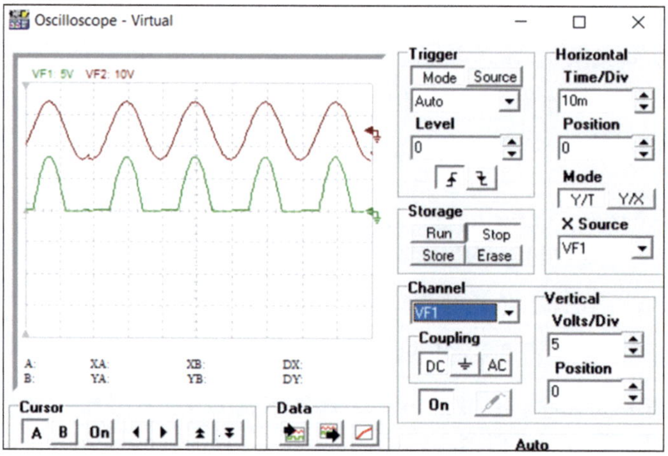

Figure 4.26 Input and output waveforms.

Now, let us connect a 5,000 μF (5 mF) capacitor in parallel with the load and plot the input-output waveforms. Figure 4.27 shows the circuit diagram and the input-output waveforms.

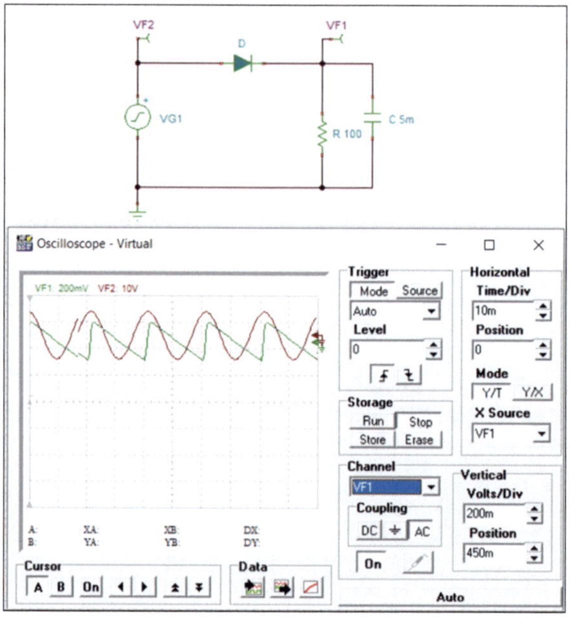

Figure 4.27 Half-wave rectifier with a load capacitor.

Full-Wave Rectifier Circuit

For efficiency reasons, rectifier circuits are often full-wave rectifiers. Such a circuit converts the bipolar input sine wave to a single polarity (usually positive) signal at its output, such that the negative pulses of the sine wave are also used instead of being discarded. Figure 4.28 shows a full-wave rectifier circuit using two diodes and a center-tapped transformer. This configuration is also known as a center-tap rectifier. When the voltage at the anode of D1 is positive, the diode conducts and current flows through the load. Similarly, when the voltage at the anode of D2 is positive, it conducts and current flows in the same direction through the load. As a result, the negative parts of the sine wave are converted to positive.

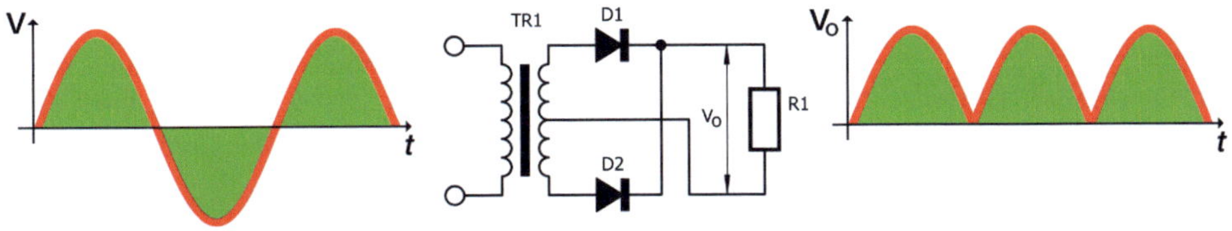

Figure 4.28 Full-wave rectifier with a center-tapped transformer.

Most full-wave rectifier circuits use a bridge configuration where four diodes are connected as shown in Figure 4.29. A bridge rectifier does not need a transformer with a center tap. The output waveform consists of positive pulses, as in Figure 4.28. When a large capacitor is used across the output to smooth the ripples, one obtains a nice DC output voltage.

During the positive cycle of the input signal, diodes D1 and D2 are forward biased and therefore conduct. D3 and D4 block and no current can flow through them. The result is a positive output pulse. During the negative cycle of the input signal, D1 and D2 block while D3 and D4 are forward biased and conduct. Again, the output is a positive pulse. Notice that the output voltage is about 1.4 V less than the applied peak voltage because of the forward voltage drop across two diodes instead of one.

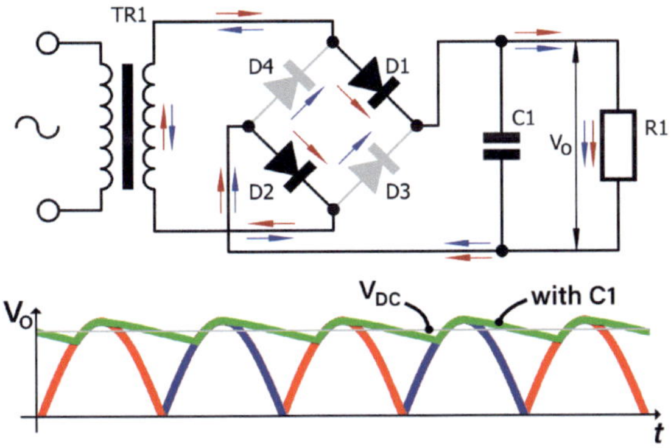

Figure 4.29 Bridge full-wave rectifier circuit. The red arrows indicate the current flow during the positive part of the input signal. D3 and D4 block in this situation. When the input signal is negative, the current (blue) flows through D3 and D4 instead of through D1 and D2. Capacitor C1 filters the output signal (green). The grey line is the average DC value.

The value of the smoothing capacitor (C1) determines the amount of ripple present in the output voltage. The bigger the capacitor value, the smaller are the ripples. Knowing the frequency of the input sine wave and the output current, we can approximately calculate the required capacitor value for the acceptable amount of ripple using the following equation:

$C = I \times T / V$

Where C is the load capacitor, I is the output current, T is the period of the rectified waveform and V is the peak output voltage. An example is given below.

Example

The input sine wave has an amplitude of 12 V with a frequency of 50 Hz. The load current is 100 mA (0.1 A). The ripple voltage must not exceed 0.5 V. Calculate the load capacitor for a full-wave rectifier circuit.

Solution

The frequency of the output signal is twice the frequency of the input signal, i.e. f = 100 Hz. Therefore, the period is, T = 1/100 = 0.01 s. Using the formula above, we obtain

$C = 0.1 \times 0.01/0.5 = 0.002\ F = 2{,}000\ \mu F$

You would use the standard value of 2,200 μF with a voltage rating of 25 V to be safe.

Bridge rectifiers are available in a single package. An example is shown in Figure 4.30.

Figure 4.30 Bridge rectifier.

Simulation

Perhaps the easiest way to understand the operation of a full-wave rectifier circuit is to simulate it and place various load capacitors across its output. The steps are as follows:

- Start the TINA-TI simulator and draw the circuit shown in Figure 4.31.
- Set a Voltage Source as the input and set it to sine wave with an amplitude of 12 V, and a frequency of 50 Hz
- Insert the monitoring points VF1 and VF2 so that input and output waveforms can be plotted. Connect a 120 Ω resistor as the load and a 2,000 μF (2 mF) load capacitor. Connect the four diodes.

- Click T&M and select oscilloscope. Set Channel to VF1, Time/Div to 10m, Volts/Div to 5. Then, set Channel to VF2 and its Volts/Div to 1. Move the input graph (green) up using the Position button. Figure 4.31 shows the input (green graph) and output (red graph) waveforms of the full-wave rectifier circuit, displayed on the oscilloscope. Notice that the ripple at the output is about 0.5 V.

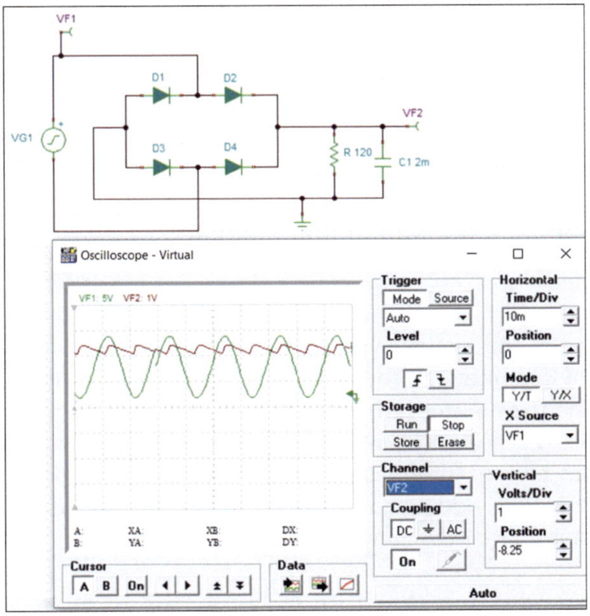

Figure 4.31 Schematic for simulating a full-wave bridge rectifier circuit.

4.3 Bipolar Transistors

The bipolar junction transistor (BJT) is a 3-terminal device. It consists of two pn-junctions formed by sandwiching either a p-type or n-type semiconductor layer between a pair of layers of the opposite type. As a result, there are two types of transistors: NPN and PNP. In an NPN transistor, two n-type layers are separated by a thin section of p-type as shown in Figure 4.32, on the left. In a PNP transistor, two p-type layers are separated by a thin layer of n-type as shown on the right in Figure 4.32.

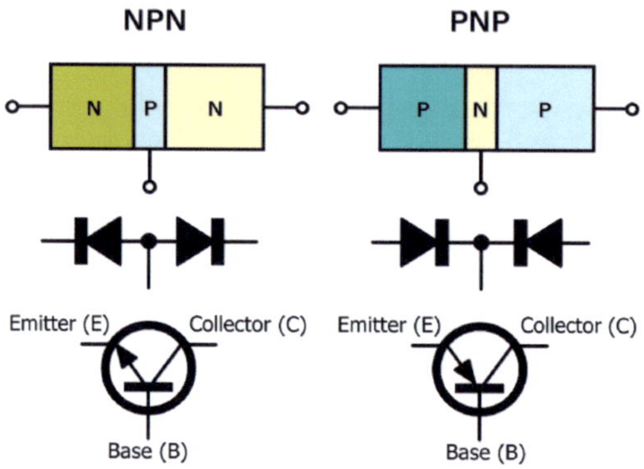

Figure 4.32 Bipolar NPN (left) and PNP (right) transistors. T he simplified construction is at the top, their diode models are in the middle and their schematic symbols are at the bottom.

Figure 4.32 also shows the circuit symbols of both NPN and PNP type transistors. The terminals are labelled E (emitter), B (base) and C (collector). The NPN transistor is identified with the arrow leaving the emitter terminal, and the PNP transistor with the arrow entering the emitter terminal. The pn-junctions can be represented by diodes, the so-called diode model, and these can be measured with a multimeter. This provides a way to identify the transistor type (NPN or PNP) and its base lead. However, the transistor's behaviour is much more complicated than that of two diodes connected in series.

4.3.1 DC Voltages and Currents in a Transistor

Figure 4.33 shows the voltages and currents in a transistor. In a NPN transistor, the base-emitter junction is forward biased by keeping the base more positive than the emitter. In a PNP transistor, the opposite is true, where the emitter is kept more positive than the base. The important equations relating to DC voltages and currents in a transistor are summarized here:

Beta (β) = I_C / I_B
Alpha (α) = I_C / I_E
$I_E = I_C + I_B$
$V_{BE} = 0.7V$

Beta is not constant, and its value changes with manufacturing tolerances and temperature. Beta is low for power transistors (up to 100) and high for small-signal transistors (up to 1000). Its minimum and maximum values are usually specified by the manufacturers.

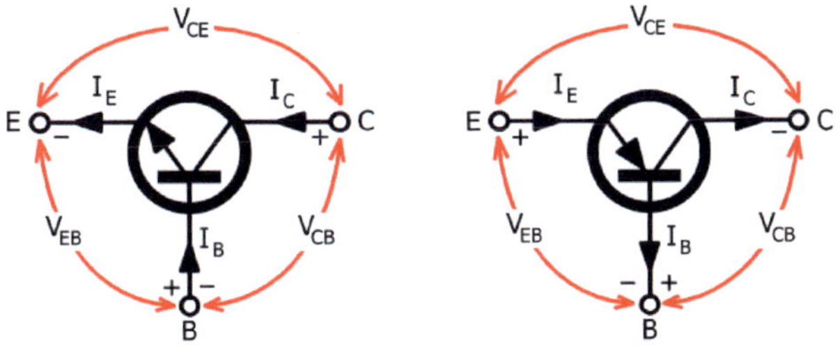

Figure 4.33 DC voltages and currents in NPN and PNP transistors.

Transistors can be used in circuits in three different configurations: common-base, common-emitter, and common-collector.

Common Base (CB) Configuration
In a common base configuration, the base terminal is common to both the input and the output (Figure 4.34 left). This is a non-inverting voltage amplifier where the input and output signals are in phase. The circuit has high voltage gain and nearly unity current gain. Because of this, it can be used as a current buffer. Also, it has a wide bandwidth, making it suitable for high-frequency applications.

Common Emitter (CE) Configuration
This is perhaps the most used bipolar transistor configuration in amplifier circuits (Figure 4.33 middle). In this configuration, the emitter terminal is common to both input and output. The voltage and current gains are medium to high, the input resistance is medium, and the output resistance is low, making it suitable to be used in most amplifier circuits.

Common Collector (CC) Configuration
In this configuration, the collector terminal is common to both input and output (Figure 4.34 right). The voltage gain of the common collector amplifier is nearly unity, the current gain is relatively high, its input resistance is high while its output resistance is low. The input and output signals are in phase. The common collector configuration is used in circuits where high input resistance is required, such as in buffer circuits.

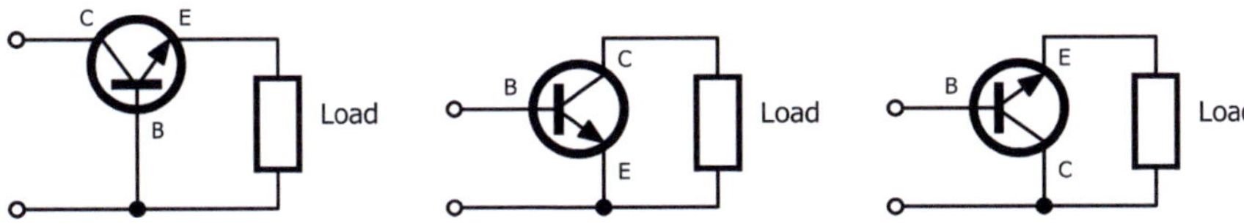

Figure 4.34 The three transistor configurations: common base (left), common emitter (middle) and common collector (right). Bias voltages and currents are not shown.

Because the common emitter is such a popular configuration, we will look at it in more detail.

4.3.2 Common Emitter Amplifier

Figure 4.35 shows a typical common-emitter-based small signal audio amplifier. We will use simple calculations to find the voltages and currents in the circuit.

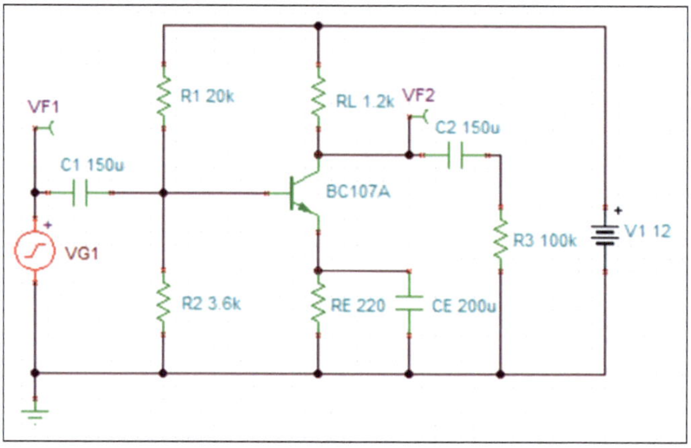

Figure 4.35 A common emitter amplifier.

We can say the following about this circuit:

- VG1 is the applied input signal.
- C1 is the input coupling capacitor. It blocks any DC component that may be present in the input signal. This is to avoid that it influences the way the transistor biased. In other words, the DC bias currents through R1, R2 and RE are not disturbed by the input signal.
- C2 is the output coupling circuit. It ensures that only an AC signal is presented to the load.
- The values of C1, C2 and CE are chosen so that their impedance is very low (almost a short circuit) for AC signals, avoiding that they affect the AC gain of the transistor.
- R1 and R2 set the base voltage VBB of the transistor. The current through R1 and R2 is usually set to be at least 10 times as large as the transistor base current. Assuming there is negligible base current (this is the case when the transistor has a high beta value, say 100 or more), the base voltage can be found from the voltage divider created by R1 and R2:
- V_{BB} = V1 × R2 / (R1 + R2) = 12 × 3600 / (3600 + 20000) = 1.83 V
- The voltage between the base and the emitter VBE is 0.7 V (the voltage over a forward-biased pn-junction, see also the section about the diode). Therefore, the emitter voltage V_{EE} is:

$V_{EE} = V_{BB} - 0.7 = 1.83 - 0.7 = 1.13$ V

- The emitter current is found to be:

$I_E = V_{EE} / RE = 1.13 / 220 = 0.0051$ A = 5.1 mA

- Ignoring the tiny base current, the collector current I_C is also 5.1 mA. Thus, the voltage drop across RL is:

$V_{RL} = 0.0051 \times 1200 = 6.12$ V

- The collector-emitter voltage V_{CE} is therefore:

$V_{CE} = V1 - V_{RL} - V_{EE} = 12 - 6.12 - 1.13 = 4.75$ V

Voltage Gain

The voltage gain of a common-emitter amplifier is given by:

$gain = g_m \times RL$

Where gm is known as the transconductance of the transistor. It is given by:

$g_m = I_C / V_T$

Furthermore,

$V_T = kT / q$

where k is the Boltzmann constant (in joules/kelvin), T is the temperature (in kelvin), and q is the electric charge (in coulombs). At 'room' temperature (16 °C), V_T = 25 mV. Therefore, the voltage gain equation becomes:

$gain = RL \times I_C / 25$

where I_C is in mA and RL is in ohms. The AC gain of our circuit (at room temperature) is now calculated using this simple formula:

$gain = V_o / V_{in} = RL \times I_C / 25$, where I_C is in mA and RL is in ohms

which gives, gain = 1200 × 5.1 / 25 = 244

This completes the approximate analysis of the DC currents and voltages in the circuit.

Simulation

We can simulate the circuit in Figure 4.35 using the TINA-TI simulator package. The steps are as follows:

- Copy the circuit from Figure 4.35. Click semiconductors and select a BC107A type NPN transistor.
- Set the value of C1 and C2 to 150 μF and CE to 200 μF.
- Configure the Voltage Generator to generate a 10 mV (0.01 V) sine wave with a frequency of 10 kHz (Figure 4.36).
- Insert monitoring points VF1 and VF2 so that we can plot the input and output voltages.
- Add a 12 V battery.
- Click T&M and select Oscilloscope.
- Set the Time/Div to 50u.
- Select Channel VF1 and set it to 10m Volts/Div.
- Select Channel VF2 and set it to 1 Volts/Div.
- Click Run to start the graphs. Move the input graph (green) higher using Position.
- Figure 4.37 shows the input (green) and output (red) waveforms. Note that the input waveform has an amplitude of 20 mV peak-to-peak; the output waveform is 3800 mV peak-to-peak. This corresponds to a voltage gain of 3800 / 20 = 190.
- Remove the Oscilloscope. Click **Analysis → DC Analysis → Table of DC results** to display the DC voltages and currents in the circuit. This is shown in Figure 4.38.
- You can easily change the values of the resistors, capacitors, battery, or the voltage generator settings and see their effects on the output response.

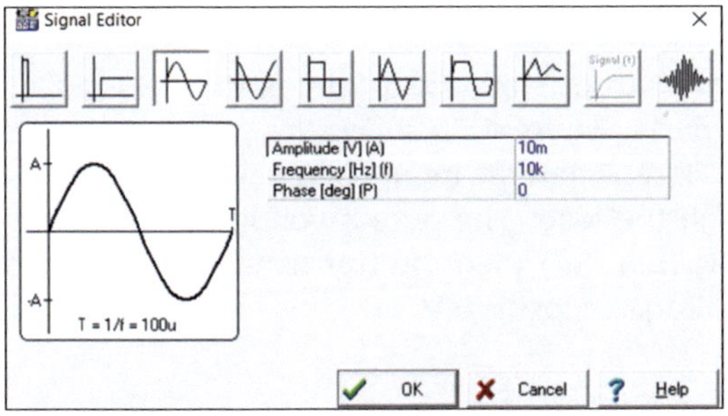

Figure 4.36 Configuring the Voltage Generator.

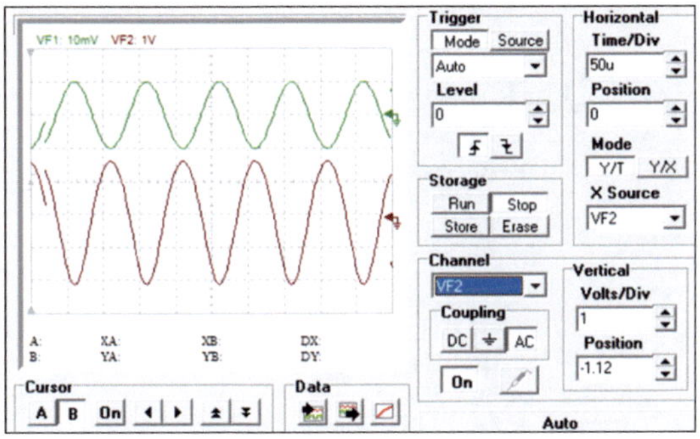

Figure 4.37 Input-output waveforms.

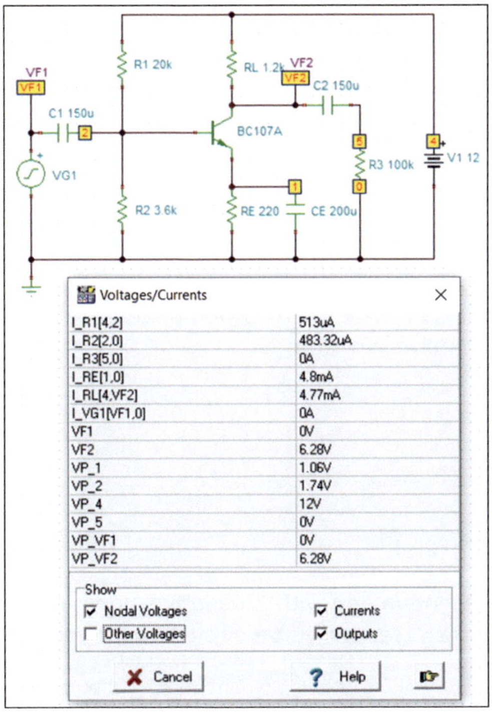

Figure 4.38 DC voltages and currents in the circuit.

4.3.3 Transistor as a Switch

One common application of transistors apart from amplifiers is as a solid-state switch, i.e. a switch without moving parts. Figure 4.39 shows the schematic of a solid-state switch made with an NPN transistor. Here, when the input voltage is LOW (e.g. 0 V) the switch is OFF and no current flows through the load. The output voltage is the same as the supply voltage. When the input is HIGH (e.g. 5V) then the transistor turns ON and current flows through the load. The output voltage drops to 0 V.

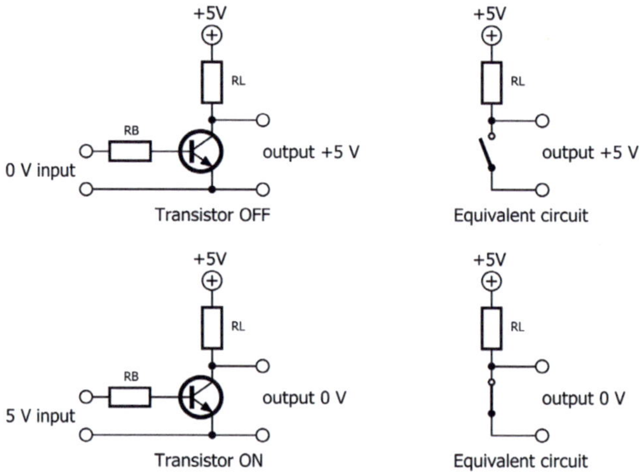

Figure 4.39 The transistor as a switch.

A PNP transistor has similar characteristics, except that the transistor is ON (aka saturated) when the input voltage is LOW, and it is OFF (aka cut-off) when the input voltage is HIGH.

Before analysing a transistor switch circuit, it is worthwhile to look at the transistor cut-off and saturation regions.

The Load Line

Looking at Figure 4.39, we can write the following relation between the transistor collector current and the collector-emitter voltage:

$$V_{CC} = I_C \times RL + V_{CE}$$

Rearranging the equation gives:

$$I_C = -V_{CE} / R_L + V_{CC} / RL$$

The above equation represents a line with a negative slope and is called the Load Line. It can be drawn on the I-V characteristic curve of a transistor, as shown in Figure 4.40.

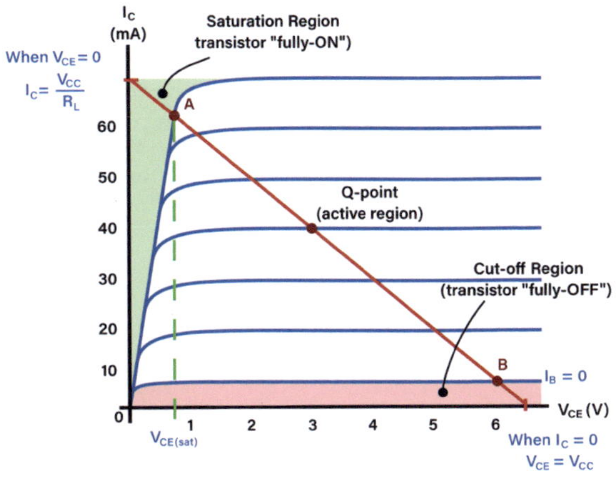

Figure 4.40 Transistor I-V curves and Load Line.

The shaded area at the bottom of the curves is the **Cut-Off** region. In this region, the transistor is OFF with no collector current (IC = 0). Also, the collector-emitter voltage (V_{CE}) is equal to the supplied voltage (V_{CC}).

The shaded region at the left of the curves is the **Saturation** region and in this region the collector-emitter voltage is nearly 0 V and there is also a large collector current. In this region, the transistor is ON with the base-emitter voltage greater than 0.7 V.

Design of a Solid-State Switch

A transistor switch can easily be designed as follows:

- Given R_L and input voltage VIN
- Assume that the collector-emitter voltage V_{CE} is 0 V
- Calculate the collector current, $I_C = V_{CC} / R_L$
- Using the minimum value of Beta (β), calculate the base current as $I_B = I_C / \beta$
- Calculate the required base resistor, $R_B = V_{IN} / I_B$

If the transistor is to switch a load with a coil such as a motor or a relay or another kind of inductive load, a diode must be connected across the load, with its cathode connected to the power supply. Its job is to protect the transistor when the latter switches the load from ON to OFF. When this happens, the magnetic field stored in the coil starts to collapse. This induces a short high-voltage pulse across the coil of opposite polarity, which may damage the transistor. The diode provides a path for this induced voltage to die away through the coil and the diode.

4.3.3.1 Project 4 – Transistor Switch With LED

Aim: The aim of this project is to use a transistor to switch an LED on and off. An LED and a current limiting resistor are connected in series as the load. Pressing a button at the base terminal turns ON the transistor and as a result, a collector current flows that turns ON the LED.

Required components:

- 330 Ω resistor
- 33 kΩ resistor
- 5 mm red LED
- BC337 NPN transistor
- Push-button switch
- Battery cable
- Breadboard
- Jumper wires
- 9 V battery

Circuit diagram: Figure 4.41 shows the schematic of the project. The pin diagram of the BC337 transistor is also shown in this figure. We will make the following assumptions:

- The transistor collector-emitter voltage V_{CE} in saturation is 0 V.
- The voltage across the LED is 1.8 V.
- The current IC through the LED is 20 mA.
- The Beta (β) of the transistor is 100 to 630.

The voltage across the current limiting resistor, $V_{RL} = 9 - 1.8 = 7.2$ V.

The value of the current limiting resistor, $R_L = V_{RL} / I_C = 7.2 / 0.020 = 360$ Ω. We will choose a 330 Ω as the nearest standard value, resulting in $I_C = 7.2 / 330 = 21.8$ mA.

The minimum Beta of the transistor is 100. The base current for saturation is therefore $I_B = I_C / β = 21.8 / 100 = 0.218$ mA.

The base voltage $V_{BB} = V_{BAT} - V_{BE} = 9 - 0.7 = 8.3$ V.

The base resistance $R_B = 8.3 / 0.218 = 38$ kΩ. We will use 33 kΩ as the nearest smaller standard value.

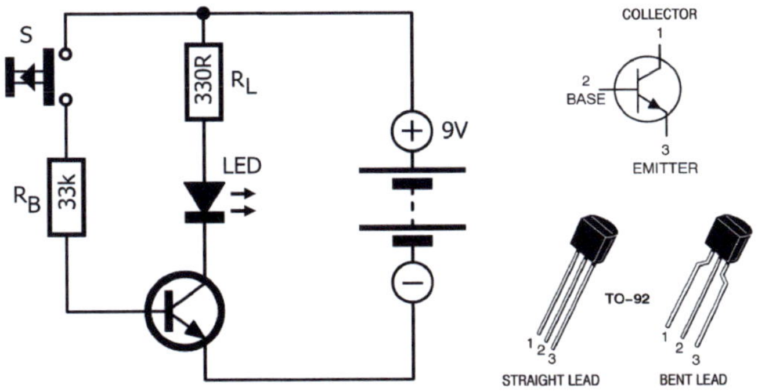

Figure 4.41 Schematic of the project.

Construction: Construct the project by placing the components on the breadboard as shown in Figure 4.42. Make sure that the transistor is connected correctly (refer to its pin diagram in Figure 4.41).

Figure 4.42 Assemble the project on a breadboard.

Procedure:

- Build the circuit.
- Connect the battery to the circuit.
- Press the button. The LED should turn ON. Release the button. The LED should turn OFF.

4.3.4 MOSFET as a Switch

A MOSFET (Metal Oxide Semiconductor Field Effect Transistor) is one of the best transistors to perform switching functions in an electronic circuit. This is a voltage-controlled transistor (a bipolar transistor is current-controlled) and it provides fast switching. It behaves like a switch in its cut-off and saturation regions of operation.

A MOSFET is a three-terminal device just like a transistor. Its terminals are labelled as Gate (G), Source (S), and Drain (D). MOSFETs come in two mode variants: depletion mode and enhancement mode.

Depletion Mode
A depletion-mode type MOSFET is normally ON. A gate-source voltage (V_{GS}) is required to switch the transistor OFF. Such a transistor behaves like a normally closed (NC) switch.

Enhancement Mode
An enhancement-mode type MOSFET is normally OFF. A gate-source voltage (VGS) is required to switch the transistor ON. Such a transistor behaves like a normally open (NO) switch.

There are four types of MOSFETs:

1. P-channel depletion mode
2. P-channel enhancement mode
3. N-channel depletion mode
4. N-channel enhancement mode

The N-channel MOSFETs are abbreviated as NMOS devices. Similarly, P-channel MOSFETs are abbreviated as PMOS devices. Figure 4.43 shows the circuit symbols of different types of MOSFETs.

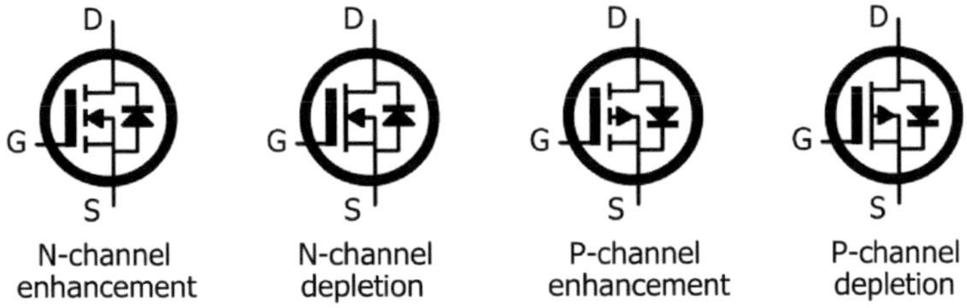

Figure 4.43 Types of MOSFETs

N-channel enhancement-mode MOSFETs are commonly used in switching applications, since the device is normally OFF when there is no gate-source voltage present.

Enhancement-Mode MOSFET as a Switch

Figure 4.44 shows N-channel and P-channel enhancement-mode MOSFETs configured to switch a load ON and OFF. In the N-channel circuit (Figure 4.44 left), a positive input gate-source voltage V_{GS} turns the MOSFET ON to make a current flow between the drain and the source. A negative or 0 V gate-source voltage turns it OFF, and no current flows between drain and source. In the P-channel circuit (Figure 4.44 right), the situation is different where 0 V turns the MOSFET ON. A positive gate-source voltage turns OFF the MOSFET and no current flows through the load. The resistors at the gates of the MOSFETs ensure that the transistors are OFF when the input is left floating.

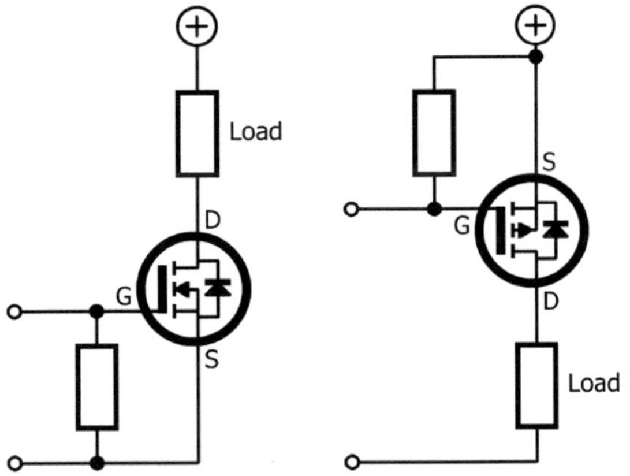

Figure 4.44 MOSFET as a switch.

The MOSFET has the following characteristics at saturation and cut-off:

Saturation

- The gate-source voltage V_{GS} is greater than the threshold voltage V_{TH}, i.e. $V_{GS} > V_{TH}$ (see below).
- The MOSFET is ON.

- Maximum drain current flows (i.e. $I_D = V_{DD} / R_L$).
- The drain voltage V_{DD} is close to 0 V.
- The MOSFET operates as a closed switch.

Cut-off

- The gate-source voltage V_{GS} is less than the threshold voltage V_{TH}, i.e. $V_{GS} < V_{TH}$ (see below).
- The MOSFET is OFF.
- No drain current flows.
- The drain voltage V_{DD} is close to the supply voltage.
- The MOSFET operates as an open switch.

MOSFETs Versus Relays

- MOSFETs consume little power when turned ON compared to relays.
- MOSFETs are fast, relays are slow.
- MOSFETs can only switch DC, while relays can switch AC as well as DC.

MOSFETs Versus Bipolar Transistors (BJTs)

- MOSFETs are voltage-controlled devices and there is no gate current. BJTs are current operated devices that require a (small) base current.
- MOSFETs can turn ON if the gate is allowed to float.
- MOSFETs are more expensive than BJTs.

The Gate Threshold Voltage (V_{TH})

The gate threshold voltage (V_{TH}) is an important parameter in MOSFET switching circuits. This parameter typically varies between 2 V and 10 V depending on the type of MOSFET.

In an N-channel enhancement-mode MOSFET, the gate-source voltage V_{GS} must be greater than V_{TH} for the MOSFET to go into saturation, i.e. ON. Similarly, in a P-channel enhancement-mode MOSFET, V_{GS} must be less than V_{TH} for the MOSFET to be ON. If, for example, the MOSFET switch is to be driven from a 5 V signal, then the V_{TH} of the MOSFET selected for the job must be less than 5 V.

If V_{TH} is higher than the control signal, then it is necessary to amplify the signal before connecting it to the MOSFET's gate. This can be done with e.g. a voltage translator or level shifter. Another possibility is to insert a bipolar transistor switch between the control signal output and the MOSFET's gate.

4.3.4.1 Project 5 – MOSFET Switch With LED

Description: The aim of this project is to use a MOSFET as a switch. An LED and a current limiting resistor are connected in series as the load. Pressing the button at the gate terminal turns ON the MOSFET and, as a result, a current starts flowing through the LED, which turns on.

Required components:

- 330 Ω resistor
- 100 kΩ resistor
- 5 mm red LED
- IRFZ44N N-channel enhancement-mode MOSFET
- Push-button switch
- Battery cable
- Breadboard
- Jumper wires
- 9 V battery

Circuit diagram: Figure 4.45 shows the schematic of the project. The pinout of the IRFZ44N MOSFET is also shown in this figure. We will make the following assumptions:

- The MOSFET's drain-source voltage V_{DS} is 0 V when ON.
- The voltage across the LED is 1.8 V.
- Current through the LED (I_{DS}) is 20 mA.

The voltage across the current limiting resistor, $V_{RL} = 9 - 1.8 = 7.2$ V

The value of the current limiting resistor, $RL = V_{RL} / I_{DS} = 7.2 / 0.020 = 360$ Ω.

We will choose a 330 Ω as the nearest value. The threshold voltage of this MOSFET is in the range from 2 V to 4 V. The gate-source voltage is 9 V, which is greater than the threshold voltage, and therefore we expect the MOSFET to be in the saturation region when the switch is pressed. Note that a gate resistor (RG) is used, since the gate of the MOSFET is floating when the switch is not pressed. The value of this resistor is not critical and can be any large value up to about 1 MΩ.

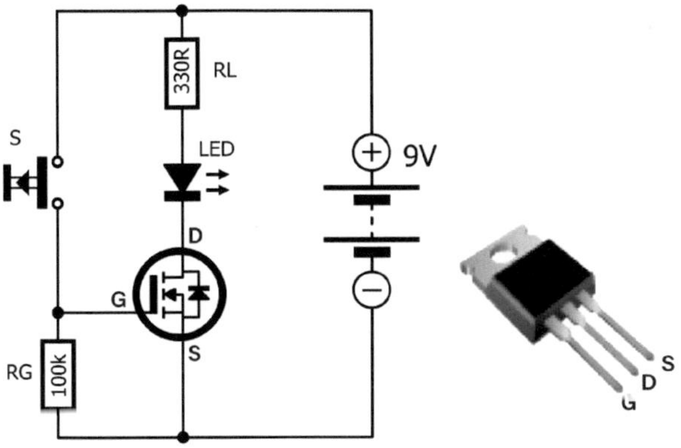

Figure 4.45 The schematic of the MOSFET LED switch.

Construction: Assemble the project by placing the components on the breadboard as shown in Figure 4.46. Make sure that the MOSFET is connected correctly (refer to its pinout in Figure 4.45)

Figure 4.46 Construct the project on a breadboard.

Procedure:

- Build the circuit.
- Connect the battery to the circuit.
- Press the button. The LED should turn ON. Release the button, the LED should turn OFF.

4.3.4.2 Project 6 – Steady Hand Game

Description: The Steady Hand Game involves trying to get a ring-shaped metal wand from one side of a wavy wire to the other side without touching it. If you happen to touch the wavy wire, then a buzzer sounds, and you have lost the game.

The game is built as a transistor switch circuit, with a buzzer connected as the load. The switch is the wand with the wavy wire. Normally, the circuit is open, and the buzzer is OFF. When you touch the wavy wire, the circuit is closed and the buzzer sounds.

Required components:

- 47 Ω resistor
- 10 kΩ resistor

- 1N4148 diode
- BC337 transistor
- Active buzzer
- Metal ring
- Wavy wire
- Battery cable
- Breadboard
- Jumper wires
- 9 V battery

Circuit diagram: Figure 4.47 shows the schematic of the project. The base circuit of the transistor is completed when the ring touches the wavy wire. This causes the active buzzer to sound.

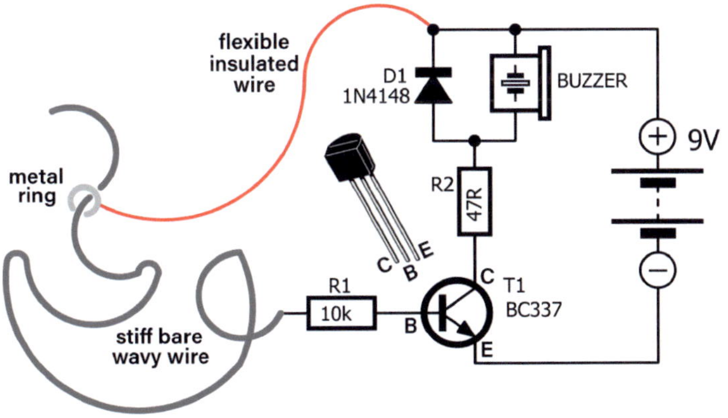

Figure 4.47 Schematic of the Steady Hand Game.

Construction: Assemble the project on the breadboard as shown in Figure 4.48. Make sure that you connect the diode and the buzzer correctly. The positive terminal of the buzzer must be connected to the power supply's positive terminal.

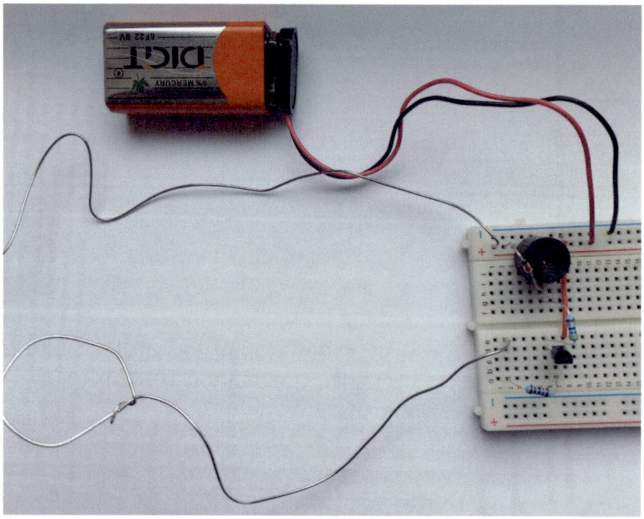

Figure 4.48 Build the project on a breadboard.

Procedure:

- Build the circuit.
- Connect the battery to the circuit.
- Touch the wavy wire with the metal ring to sound the buzzer. Play the game yourself and let your friends and/or family members play it as well.

Chapter 5 • Simple Circuits Based on Previous Chapters

5.1 Overview

In this chapter, we will learn the operation and use of some other electronic components in more detail. Here, we will cover the mechanical switches, relays, oscillator circuits, optocouplers, and the 555 timer circuits.

5.2 Mechanical Switches

A mechanical switch is an electro-mechanical device which is used to open (break) or close (make) the electrical path in a circuit. Switches are found in almost every electrical circuit to turn the power on or off but also to (de)activate functions on equipment or to enter information.

The most frequently used type of switch is a mechanical one that includes a set of movable contacts. By moving an arm these contacts can either be made to touch each other or be separated.

Switches can have momentary or latching action. Momentary switches (e.g. push-buttons, computer keyboard keys, etc.) only make contact while the switch is being pressed. Latching switches (e.g. the kitchen light switch) stay in their last position until it is changed to another position. They have memory.

Mechanical switches are classified by the number of **poles** and **throws** they have. The number of **poles** of a switch defines how many separate circuits it can control. Therefore, a switch with one pole can only control a single circuit, a double-pole switch can control two. The number of **throws** on a switch defines how many positions each of the switch's poles can be connected to. For example, if a switch has two throws, each pole in the switch can be connected to one of two terminals. The following list shows how switches are classified according to their number of poles and throws:

- Single-Pole Single-Throw (SPST)
- Single-Pole Double-Throw (SPDT)
- Double-Pole Single-Throw (DPST)
- Double-Pole Double-Throw (DPDT)

When a switch has more than two poles and/or throws, the number is indicated by a digit, e.g. SP6T or 3P3T.

Additionally, the following mechanical switches are commonly used:

- Push-button switch
- Limit switch
- Float switch
- Rotary switch

SPST Switch

This is a basic ON-OFF type of switch that makes or breaks a single contact. The SPST switch has two terminals. Figure 5.1 shows the use of an SPST switch in an LED circuit.

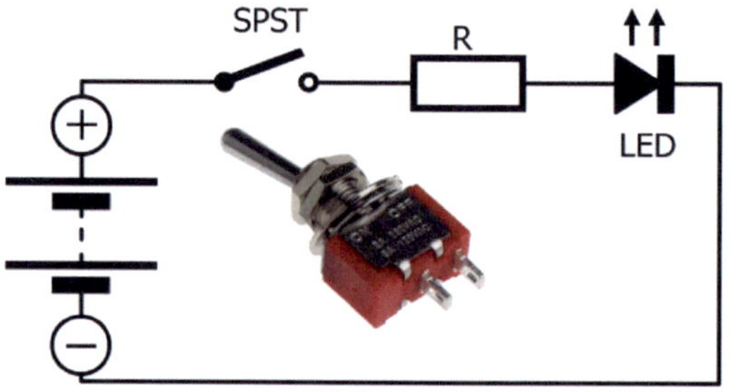

Figure 5.1 An SPST switch can control an LED.

SPDT Switch

The SPDT switch has three terminals. As the switch has two ON positions it is sometimes also called ON-ON switch. This kind of switch is commonly used as changeover switch to toggle an input between two outputs (or two outputs with one input). There exist also SPDT switches with an OFF position in between the two ON positions. These are known as ON-OFF-ON types. Figure 5.2b shows the use of an SPDT switch in an LED circuit.

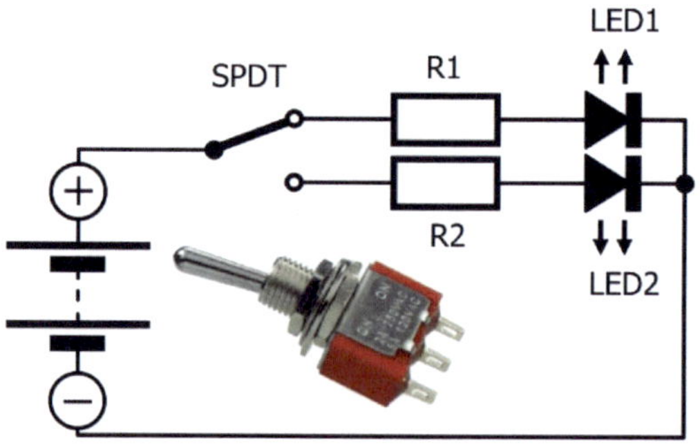

Figure 5.2 SPDT switch to toggle between two LEDs.

DPST Switch

DPST switch has four terminals, two input contacts and two output contacts. A DPST switch behaves like two separate SPST switches, but operating at the same time. The switch has only one ON position. This type of switch is commonly used to control two different circuits at the same time or to disconnect an appliance from the mains. Figure 5.3 shows the use of a DPST switch in a basic LED circuit.

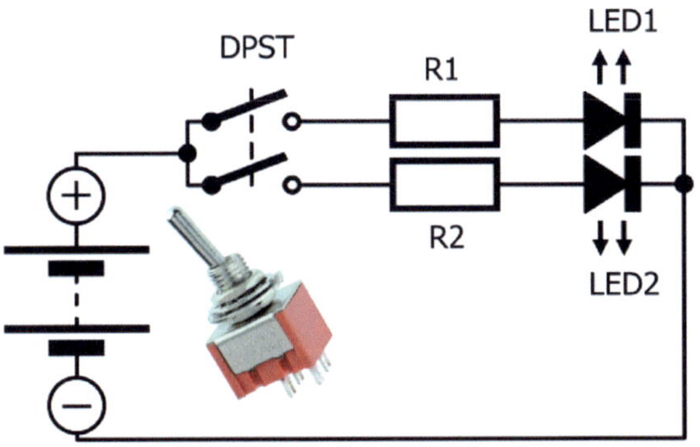

Figure 5.3 DPST switching two LEDs at the same time.

DPDT Switch

The DPDT switch has six terminals, two are input contacts while the remaining four are the output contacts. This switch behaves like two separate SPDT switches, operating at the same time. This type of switch is commonly used to control either two different circuits at the same time, or different functions of a circuit, such as the speed and direction control of a DC motor. Figure 5.4 shows the use of a DPDT switch in an LED circuit. Toggling the switch in one direction turns ON LED1 and LED3. Toggling it in the other direction turns on LED2 and LED4. DPDT switches also come in ON-OFF-ON configurations.

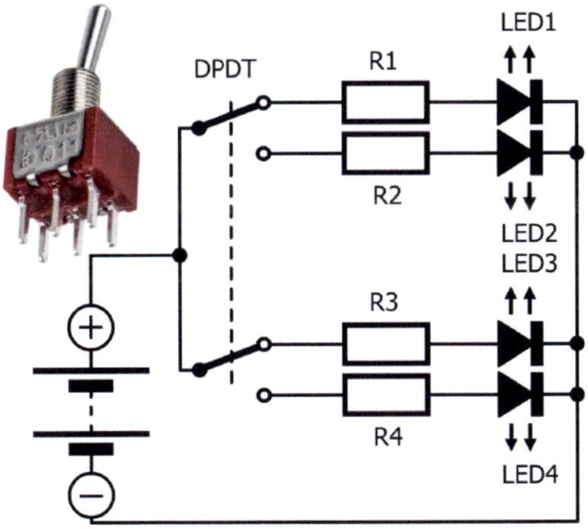

Figure 5.4 DPDT switch controlling four LEDs.

Push-button Switch

This is a momentary contact switch that makes a contact as long as the switch is pressed (see Figure 5.5). A spring is used to return the switch back to its original state when released. Push button switch contacts are generally normally open and are closed when pressed.

Latching push-button switches exist too and are commonly encountered as power on/off buttons on e.g. audio equipment and appliances. In this book, when a push-button is mentioned, we refer to the momentary type.

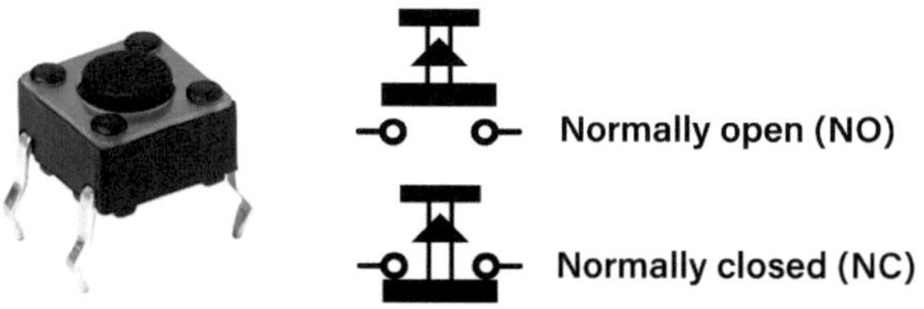

Figure 5.5 Push-button switch. The shown type is also known as a tactile switch. These switches are often mounted hidden behind a soft or flexible surface.

Limit Switch

Limit switches (Figure 5.6) are momentary switches operated by the motion of an object. The object pushes or releases the lever of the switch, thus changing its state. A spring ensures that the switch returns to its idle state when the object is removed. Limit switches are often 3-terminal SPDT switches. They are found in e.g. household appliances to detect if a door is open or closed or in CNC machines to detect if the spindle motor has reached the extremity of an axis.

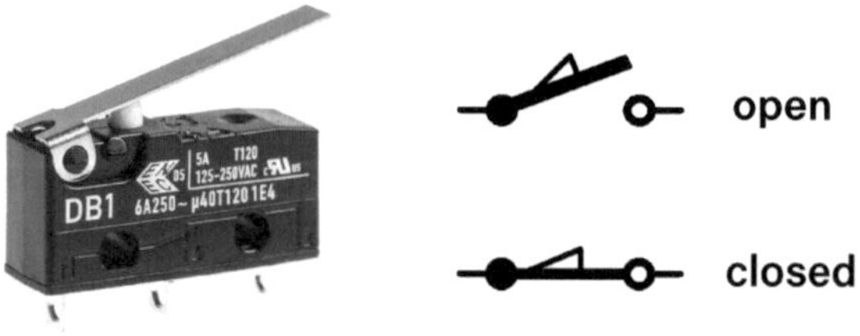

Figure 5.6 Limit switch.

Float Switch

Float switches (Figure 5.7) are momentary switches operated by a liquid or fluid like water or oil. When a rising liquid level activates the switch, a pump can turn on or off or an alarm can sound. Similarly, when a falling liquid level deactivates the switch, a pump can turn on or off or an alarm can sound. This allows for instance controlling the level of a fluid in a tank.

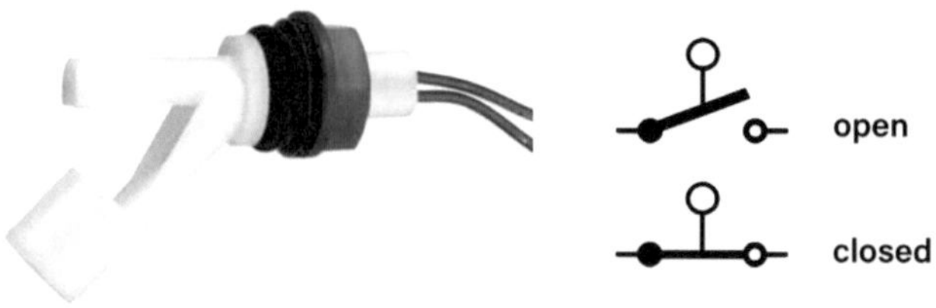

Figure 5.7 Float switch.

Rotary Switch

Rotary switches are used to connect one line to many lines (Figure 5.8). One common application of rotary switches is in range selector circuits, for example selecting the bands in a radio, or setting the measuring function and range of a multimeter. Rotary switches come in many different arrangements, such as 3-pole 4-way, 10-way, 12-way, etc.

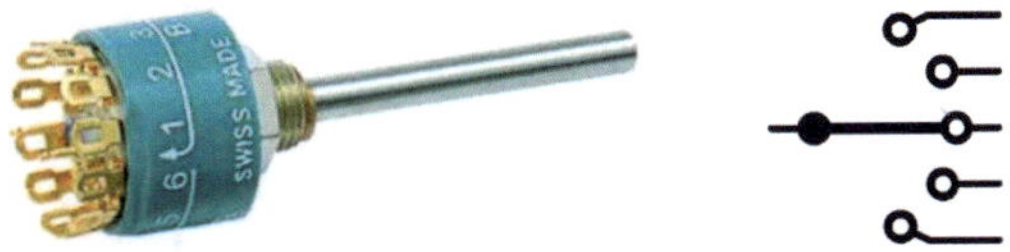

Figure 5.8 Rotary switch. The model shown is a 2-pole 6-way type. Some rotary switches feature an adjustable end-stop (a removable ring under the nut) allowing the same model to be used in different configurations.

5.2.1 Project 1 – Upstairs-Downstairs Two-Way Switch

Aim: You have probably come across two-way switches used in two storey houses. Here, the light is controlled by a switch located downstairs and by another one upstairs. The same light can be turned ON and OFF with either switch. For example, the light can be turned ON using the downstairs light and then it can be turned OFF using the upstairs light, and vice-versa. In this project, an LED is used with a 330 Ω current limiting resistor.

Required components:

- 2× SPDT switch
- 5 mm red LED
- 330 Ω resistor
- Battery cable
- Breadboard
- Jumper wires
- 9 V battery

Circuit diagram: Figure 5.9 shows the schematic of the project. SPDT switches have 3 pins where the middle pin is the common (C) pin.

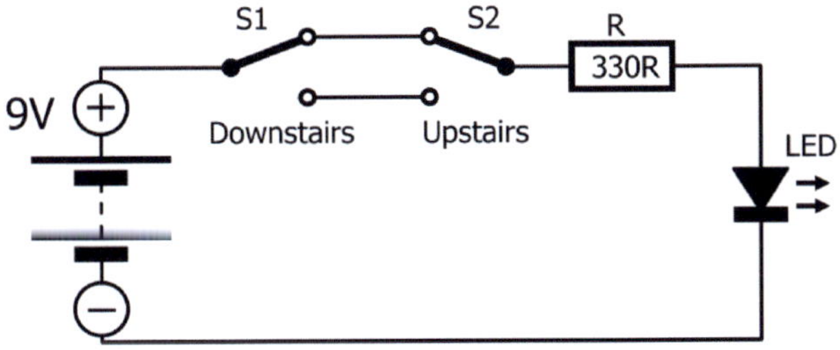

Figure 5.9 Schematic of the upstairs/downstairs LED circuit.

Construction: Assemble the project by placing the components on the breadboard as shown in Figure 5.9.

Procedure:

- Build the circuit.
- Connect the battery to the circuit.
- If the LED is ON, toggle any of the switches to turn OFF the LED. Similarly, if the LED is OFF, toggle any of the switches to turn it ON.

5.2.2 Contact Bounce Problems

When a mechanical switch is closed, the hard metal contacts will bounce against each other. The effect is identical to a (rubber) ball falling on the floor. The ball will bounce a few times before it lies still. The same is true for mechanical switch contacts. It can take as long as a few tens of milliseconds before the contact becomes stable. Similarly, when a switch is opened, the contacts scrape and bounce against each other a bit before they separate, resulting in intermittent contact. These effects are known as contact bounce. All mechanical switches suffer from contact bounce.

Contact bounce is a problem in high-voltage and high-current circuits where the bouncing contacts may create sparks. These sparks can result in burned contacts (switch always open) or even weld contacts together (switch always closed). Contact bounce is also a problem in fast systems like microcontroller circuits that can read the state of a switch several times per millisecond. When switch contacts bounce, a fast system may interpret each bounce as a separate keypress and execute the corresponding function every bounce instead of just once. Several contact debouncing techniques are available to minimize or avoid problems caused by contact bounce.

5.3 Project 2 – Relay With LED

Aim: This is a simple relay-based project. In this project an LED is turned ON by a relay when a button is pressed. The aim of the circuit is to show how a relay can be used in a simple circuit.

Required components:

- 9 V SPDT relay (electromechanical or SSD)
- 5 mm red LED
- 330 Ω resistor
- Push-button
- 2× Battery cable
- Breadboard
- Jumper wires
- 2× 9 V battery

Circuit diagram: Figure 5.10 shows the schematic of the project. A current limiting resistor is used in series with the LED. Pressing the button closes the circuit and activates the relay. The relay in turn enables current to pass through the LED, which turns ON.

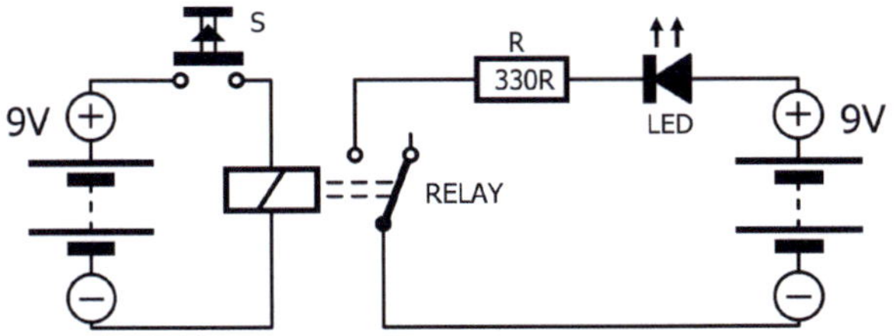

Figure 5.10 Circuit diagram.

Construction: Assemble the project by placing the components on the breadboard as shown in Figure 5.10.

Procedure:

- Build the circuit.
- Connect the batteries to the circuit.
- If the LED is ON, use the other terminal of the relay. Press the button. The LED should turn ON. If it is an electromechanical relay, you should hear the relay contacts clicking. Release the button, the LED should turn OFF.

5.4 Electronic Buzzers

Buzzers are audio signalling devices, which are used in many applications to provide sound and audible feedback such as alarms, beeps, and melodies. They are small devices (Figure 5.11), usually operated from around 3 to 5 volts but more powerful devices exist too.

Figure 5.11 A buzzer. The '+' symbol indicates its polarity, which is especially important for active buzzers. Often, the little hole in the buzzer is covered by a seal. This protects the buzzer against liquids used in automated board assembly processes. The seal must be removed after assembly to obtain the best and loudest sound.

Most buzzers nowadays are piezoelectric, consisting of two electrical contacts placed on both faces of a small disk of piezoelectric material. When a voltage is applied across the two terminals, the piezoelectric material deforms. Doing this repeatedly creates sound, like a loudspeaker.

There are basically two types of buzzers: active and passive.

Active Buzzers
Active buzzers produce sound when they are connected to a DC power supply (e.g. a battery). They can produce a single tone, which is tuned to around 2 kHz by most of the manufacturers. Active buzzers contain a tiny oscillator circuit that drives the piezo element. Therefore, they are polarized, i.e. they have a positive and a negative terminal. The positive terminal is usually marked with a plus ('+') sign on the body of the buzzer or near one of the terminals. Connecting an active buzzer the wrong way around may break the device.

Passive Buzzers
Passive buzzers are like loudspeakers and need an AC signal to produce sound. When connected to a DC supply, they remain silent. Depending on the model, they can reproduce sounds covering almost the complete audio range (40 Hz to 20 kHz) but there are also types optimised for ultrasonic sounds. Microcontroller systems often use buzzers to play square wave sounds.

In general, active and passive buzzers cannot be distinguished from each other by simply looking at them. One way to identify them is to connect the buzzer to a battery by observing the correct polarity. If it is an active buzzer, then a loud tone will be generated. If, on the other hand, it is a passive buzzer, then a pop sound will be generated.

5.5 Optocouplers

An optocoupler (also called an optoisolator) is an integrated circuit (IC) that allows an electrical signal to be transmitted between two circuits using light waves. Optocouplers provide galvanic isolation between their inputs and outputs.

An optocoupler incorporates a transmitter and a receiver. The transmitter consists of an LED that emits light (usually infrared light), and the receiver is a photosensitive device (usually a transistor base) that detects the emitted light. Figure 5.12 shows the internal structure of a typical optocoupler and a commercially available optocoupler chip.

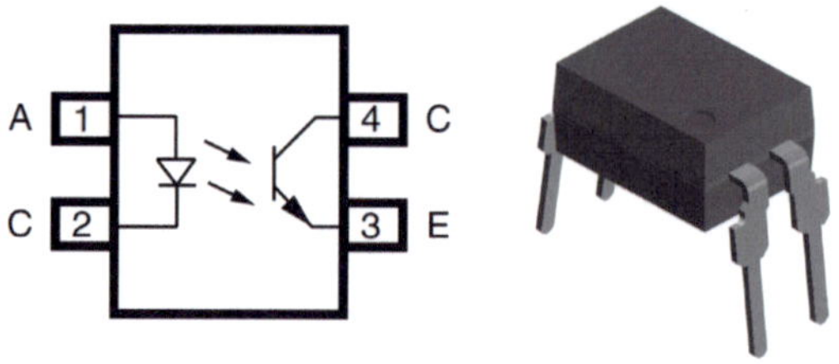

Figure 5.12 An optocoupler

Optocouplers tend to be rather large devices. The reason for this is the galvanic isolation they provide. For proper galvanic isolation, the input and output pins may not be close together. Therefore, their packages are relatively large. When the galvanic isolation qualities of an optocoupler are required for an application (e.g. in certain switch-mode power supplies), it is important not to route any signals traces under it. If you do this anyway, the safety distance between the two isolated circuits is no longer respected and the specified galvanic protection level can no longer be met.

5.5.1 Example Optocoupler Circuit

Figure 5.13 shows a simple optocoupler circuit. In this circuit, LED1 is connected to the output of the optocoupler device. When the button is pressed, the LED inside the optocoupler emits light, which is sensed by its detector. This causes current to flow in the output circuit that contains the LED1. As a result, LED1 is turned ON. Note that a transistor-based optocoupler is used in this circuit.

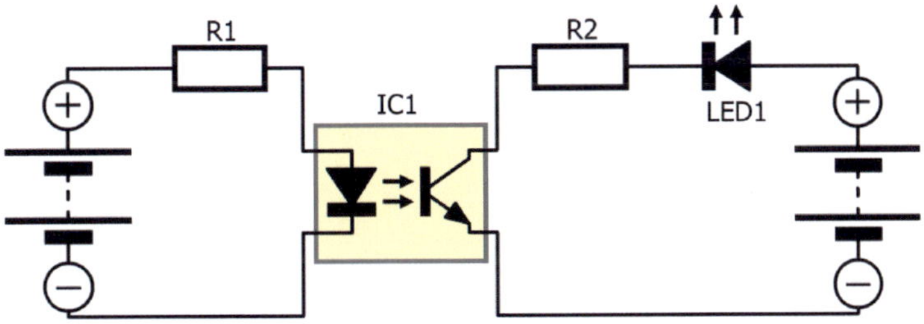

Figure 5.13 Simple optocoupler circuit.

5.6 Transistor Oscillators and Other Circuits

Oscillators are electronic circuits that produce periodically repeating signals. They are found in so many electronic circuits that it is probably easier to list those that do not use them, such as audio amplifiers.

Depending on the design, oscillators can generate sine, saw, triangle, pulse, square, and rectangle wave signals. Oscillators in radio circuits usually produce sine waves, while square wave oscillators are used in microcontroller systems and in most other digital circuits.

In this section, we will look at the properties and design equations of transistor-based sine wave and square wave oscillators and simulate them using the TINA-TI circuit simulator. The advantage of using a simulator is that you can change the component values or the design easily without having to build any circuits and then see their effects on the output response.

5.6.1 Sine Wave Transistor Oscillators

Every sine wave oscillator satisfies two conditions:

1. The output of the amplifier is fed back to the input in such a way that the overall sum of the amplifier and feedback network phase shift equals 0° (or 360°).

2. The overall amplifier gain including the feedback network is equal to 1. If the overall gain is less than 1, then the circuit will not oscillate. If, on the other hand, the overall gain is greater than 1, then the output waveform will be distorted.

Colpitts Oscillator

This is a very popular oscillator design used mainly in high-frequency radio circuits. The Colpitts oscillator uses a capacitive voltage divider network as its feedback source. Figure 5.14 shows the circuit diagram of a bipolar-transistor-based typical Colpitts oscillator.

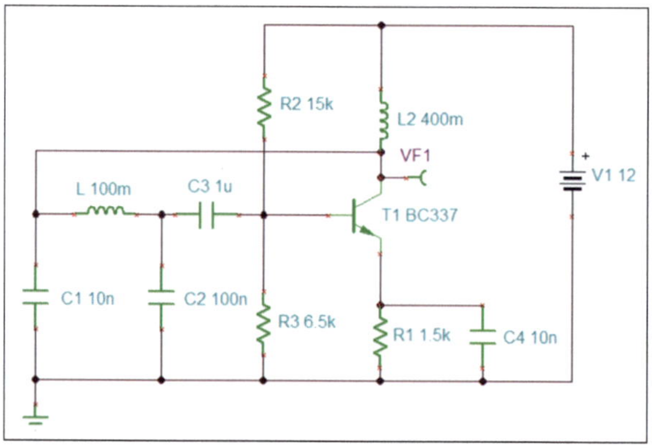

Figure 5.14 Colpitts oscillator. The oscillation frequency is determined by L, C1 and C2. C3 blocks the DC current so that the bias of the transistor is not disturbed by the feedback network.
L2 decouples the output from the power supply.

The frequency is set by the two capacitors C1 and C2 and inductor L. The oscillation frequency is given by:

$$f = \frac{1}{2\pi\sqrt{LC_S}}$$

Where f is the frequency in hertz, L is the inductance in henries, and C_S is the series capacitance of C1 and C2 in farads, i.e.

$$C_S = \frac{C1 \times C2}{C1 + C2}$$

The amount of feedback in the circuit, the feedback ration is C1:C2.

On the Internet, you can find free calculators for calculating the oscillation frequency of a Colpitts oscillator given the values of the two capacitors and the inductor.

Simulation

Let us simulate the Colpitts oscillator shown in Figure 5.15. From the theory, we expect the oscillation frequency to be:

$$f = \frac{1}{2\pi\sqrt{100 \times 10^{-3} \times 9.09 \times 10^{-9}}}$$

Which gives, f = 5.28 kHz (i.e., a period of 189 μs). The feedback ratio C1:C2 = 10%.

- Draw the circuit as shown in Figure 5.14.
- Choose a BC337 transistor and set the battery voltage to 12 V.
- Insert the monitoring point VF1 at the collector of the transistor.
- Click T&M and select the Oscilloscope.
- Set Time/Div to 100u, Volts/Div to 5, Coupling to AC.
- Click Run to start the oscilloscope. Figure 5.16 shows the oscillator wafeform at the collector of the transistor.

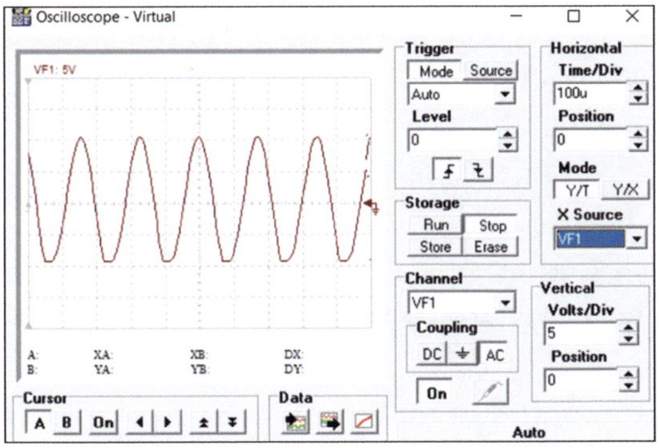

Figure 5.15 Output of the transistor.

Advantages of the Colpitts oscillator:

- The oscillation frequency can be very high.
- The frequency is stable.
- The frequency can be varied by varying the capacitors.
- The frequency is not affected by temperature changes.
- The design is simple, requiring only two precision capacitors and an inductor.

Hartley Oscillator

The Hartley oscillator is similar to the Colpitts oscillator, but it uses two indictors and a capacitor in the feedback network instead of two capacitors and an inductor. Figure 5.16 shows an example of a Hartley oscillator. The middle part of the two inductors is grounded. One side of the capacitor is connected to the base of the transistor while the other side is connected to the collector. Capacitors C2 and C3 are coupling capacitors that block the DC current so that the bias of the transistor is not disturbed by the feedback network.

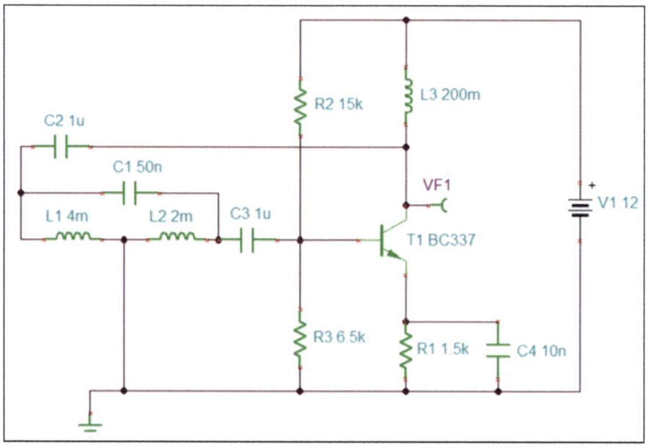

Figure 5.16 Hartley oscillator.

The oscillation frequency depends on the values of the two inductors and the capacitor in the feedback network and is given by:

$$f=\frac{1}{2\pi\sqrt{C\,L_T}}$$

Where, *f* is the frequency in hertz, C is the capacitance in farads, and L_T is the total inductance in henries. L_T is given by the series connection of inductors L1 and L2. i.e.

L_T = L1 + L2

As an example, suppose L1 = 4 mH, L2 = 2 mH, and C = 50 nF. The total inductance is L_T = L1 + L2 = 6 mH. The oscillation frequency is given by:

$$f=\frac{1}{2\pi\sqrt{50\times10^{-9}\times6\times10^{-3}}}$$

Which gives, *f* = 9.19 kHz.

You can use freely available programs on the Internet to calculate the oscillation frequency of the Hartley oscillator given the values of the two inductors and the capacitor.

Phase-Shift Oscillator

A phase-shift oscillator is also known as an RC oscillator. In such a circuit the feedback network is built with resistors (R) and capacitors (C) without inductors (L). Oscillators such as the Colpitts and Hartley oscillators are LC-type circuits. In general, LC oscillators have the following drawbacks (note that Colpitts and Hartley oscillators were designed to overcome some of these drawback):

- LC oscillators cannot be used for low frequencies.
- Inductors can be bulky and expensive.
- The waveform can be poor unless it is designed carefully.
- The frequency stability is not so good.

Using only resistors and capacitors in the oscillator circuit eliminates the above-mentioned drawbacks. Phase-shift oscillators generate good sine waves with excellent stability if precision resistors and capacitors are used. Perhaps the biggest advantage of the phase-shift oscillators is that it can be used for (very) low frequencies. However, one of its main drawbacks is that it may be difficult to start the oscillation.

Figure 5.17 shows the circuit diagram of a phase-shift oscillator. The number of RC networks used determines the number of stages. In the circuit in Figure 5.21, a 3-stage phase-shift oscillator is shown.

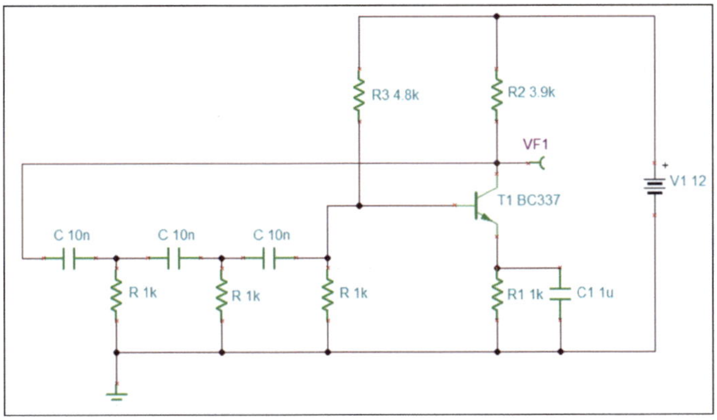

Figure 5.17 3-stage phase shift oscillator.

The oscillation frequency is given by:

$$f = \frac{1}{2\pi RC\sqrt{2N}}$$

Where *f* is the frequency in hertz, R is the resistance in ohms, C is the capacitance in farads, and *N* is the number of stages used in the feedback network.

The oscillation frequency in Figure 5.21 is calculated as follows:

$$f = \frac{1}{2\pi \times 10^3 \times 10 \times 10^{-9} \times \sqrt{6}}$$

Which gives f = 6.5 kHz.

5.6.2 The Multivibrator

A multivibrator is a two-stage resistor-capacitor coupled amplifier with positive feedback from each of the transistors to the other one. The circuit generates rectangle waves (including square waves). Multivibrators are commonly used as signal generators, time delays, frequency dividers, memory elements, and so on.

There are three types of multivibrators (Figure 5.18): astable, monostable, and bistable. An astable multivibrator is a free-running oscillator. The frequency of the generated waveform is set by the values of resistors and capacitors used in the circuit.

A monostable multivibrator (also called a one-shot multivibrator) has a trigger input (T). The circuit produces a single output pulse for each trigger pulse. The duration of the output pulse is set by resistors and capacitors. Monostable multivibrators are commonly used in time delay and wave shaping circuits.

A bistable multivibrator (also called a flip-flop) has two stable states, and it requires two trigger pulses (T1 and T2, or Set and Reset) to toggle between these states. Applying trigger pulse T1 changes the state. Applying trigger pulse T2 changes the state back to its previous state.

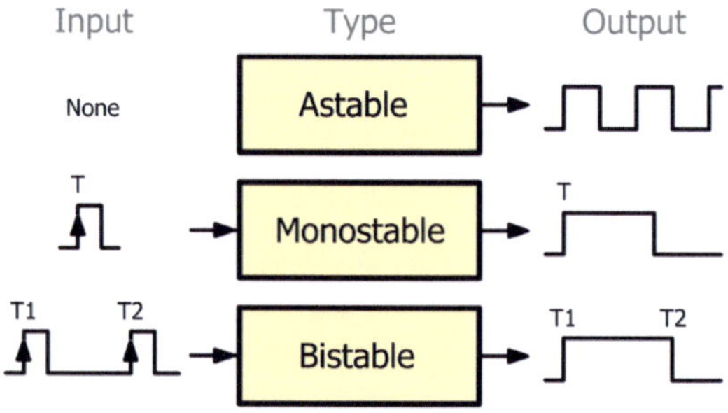

Figure 5.18 Types of multivibrators.

Because multivibrators are used in many applications, we will examine them in more detail below.

The Astable Multivibrator

As mentioned earlier, the astable multivibrator is a free-running oscillator. Figure 5.19 shows a multivibrator circuit. Two transistors are cross-coupled, and feedback is applied between them by means of the capacitors.

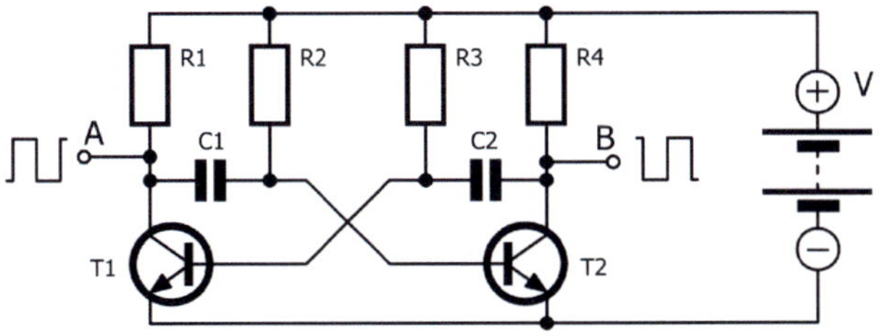

Figure 5.19 Astable multivibrator.

The oscillation frequency of the astable multivibrator is set by capacitors C1 and C2, and resistors R2 and R3:

Output ON time (Mark) = 0.69 × C1 × R3
Output OFF time (Space) = 0.69 × C2 × R2
Duty cycle (%) = 100 × Mark / (Mark + Space)

The mark-to-space ratio (and therefore the duty cycle) and the frequency can easily be changed by altering the values of the Rs and/or Cs in the equations above. Often, C1 = C2 (= C) and R2 = R3 (= R) which corresponds to a symmetrical output where Mark and Space are equal. The duty cycle in that case is 50%. The oscillation frequency is given by:

$$f = \frac{1}{1.38 \times R \times C}$$

Where R is in ohms, C is in farads and the frequency is in hertz. Table 5.2 lists some values for R and C and the resulting frequency. Only some standard-value resistors and capacitors are shown in the table.

Resistor (kΩ)	Capacitor (nF)				
	1	4.7	22	100	470
1	724	154.2	32.9	7.24	1.54
4.7	154	32.7	7	1.54	0.327
22	32.9	7	1.49	0.329	0.070
47	15.4	3.28	0.700	0.154	0.0328
100	7.24	1.54	0.329	0.0724	0.0154
470	1.54	0.328	0.070	0.0154	0.0328
1,000	0.724	0.154	0.0329	0.072	0.0154

Table 5.2 Astable multivibrator frequency table. The frequency is given in kHz.

5.6.2.1 Project 3 – Astable Multivibrator Flashing LEDs

Aim: This project shows how two LEDs can be connected to a multivibrator circuit to make them flash alternatingly.

Required components:

- 2× BC337 NPN transistor
- 2× 100 kΩ resistor
- 2× 1 kΩ resistor
- 2× 10 µF capacitor
- 2× 5 mm red LED
- Push-button
- Battery cable
- Breadboard
- Jumper wires
- 9 V battery

Circuit diagram: Figure 5.20 shows the schematic of the project. The LEDs are connected in series with the collector of the transistors through 1 kΩ current limiting resistors. Assuming the collector-emitter voltage is 0 V at saturation, the current through the LEDs is:

> I = (9 – 1.8) / 1000 = 7.2 mA. Smaller resistors (e.g. 470 Ω) can be used for higher LED currents.

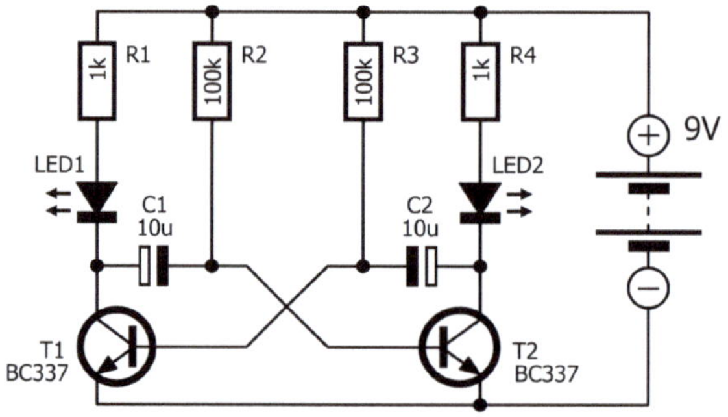

Figure 5.20 Schematic of the alternating LED blinker.

With the chosen components, R = 100 kΩ and C = 10 µF, the oscillation frequency is:

$$f = \frac{1}{1.38 \times 100 \times 10^{3} \times 10 \times 10^{-6}}$$

Which gives f = 0.725 Hz. Therefore, the period is 1.38 seconds, giving 0.69 s ON time and 0.69 s OFF time. The duty cycle is 50%. Figure 5.21 shows the output waveform.

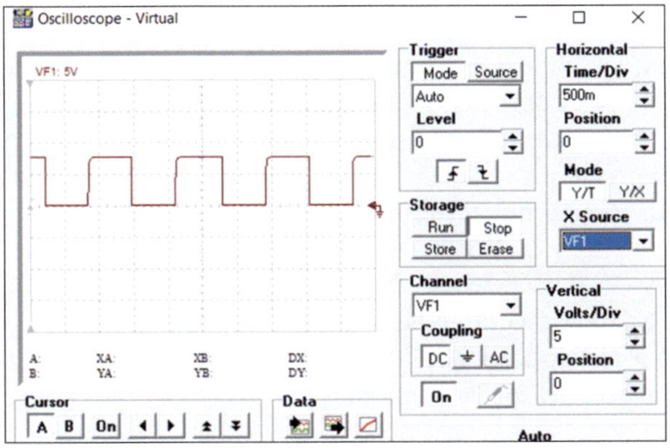

Figure 5.21 Output waveform of the multivibrator.

Construction: Assemble the project by placing the components on the breadboard (Figure 5.22).

Figure 5.22 The multivibrator built on a breadboard.

Procedure:

- Build the circuit.
- Connect the battery to the circuit.
- The two LEDs should flash alternately at a rate of about just over half a second each.

Note: the oscillation frequency can be made adjustable while keeping the output waveform symmetric by using a dual-ganged potentiometer to replace resistors R2 and R3.

The Monostable Multivibrator

As mentioned earlier, a monostable multivibrator needs to be triggered by an external pulse. Once triggered, the output changes state (e.g. goes from 0 to 9 V) for a while and then returns to 0 V. The duration of the pulse is set by a resistor and capacitor. An example project is given below, where the LED is turned on for 7 seconds when a button is pressed. Here, the button is used to provide the trigger signal to the circuit.

5.6.2.2 Project 4 – Monostable Multivibrator Timed LED

Aim: In this circuit, an LED is connected to the output of a monostable multivibrator and a push-button is used as a trigger input. Pressing the button creates a trigger signal and as a result the LED is turned ON for 7 seconds.

Required components:

- 2× BC337 NPN transistors
- 22 kΩ resistor
- 2× 1 kΩ resistor
- 10 kΩ resistor
- 10 μF capacitor
- 470 μF capacitor
- 5 mm red LED
- Push-button switch
- Battery cable
- Breadboard
- Jumper wires
- 9 V battery

Circuit diagram: Figure 5.23 shows the schematic of the project. The LED is connected to the output of the multivibrator. Pressing the button will create a trigger signal where the LED will turn ON for 7 seconds. The duration of the output pulse (LED ON time) is given by:

$$T = 0.69 \times R2 \times C1$$

In this example, R2 = 22 kΩ and C1 = 470 μF. Using the above formula gives the pulse duration as:

$$T = 0.69 \times 22 \times 10^{3} \times 470 \times 10^{-6}$$

Which gives, T = 7.1 seconds.

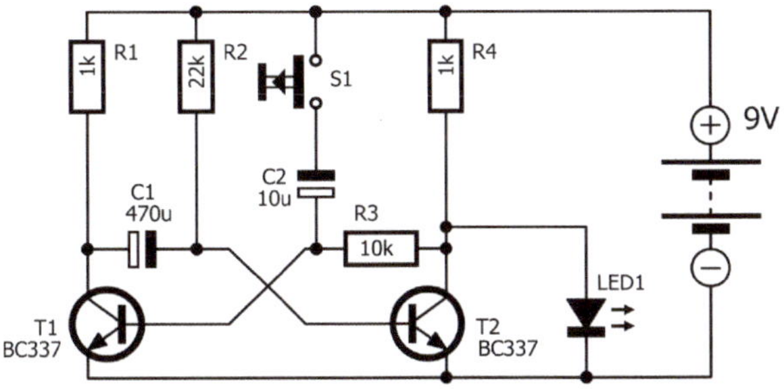

Figure 5.23 Monostable multivibrator circuit.

Construction: Assemble the project by placing the components on the breadboard (Figure 5.24).

Figure 5.24 The monostable multivibrator built on a breadboard.

Procedure:

- Build the circuit.
- Connect the battery to the circuit.
- Press the push-button. The LED should turn ON for 7 seconds and then turn OFF.

5.7 The 555 Timer

We have seen above that free-running oscillators, and monostable and bistable multivibrators can be built with a pair of transistors and some resistors and capacitors. Because multivibrators are so common in electronic circuits, a dedicated IC was designed to carry out these tasks. Originally launched in 1972 under the name NE555, today this device is still one of the most popular ICs around. The '555' is cheap and has been used in all sorts of applications, from simple timers to class-D amplifiers. It can be a timer to generate a single pulse with a well-defined duration, but it can also be a free-running oscillator with varying duty cycles.

The 555 timer is an 8-pin device, and it is extremely robust and easy to use. Figure 5.25 shows the chip together with its internal simplified block diagram.

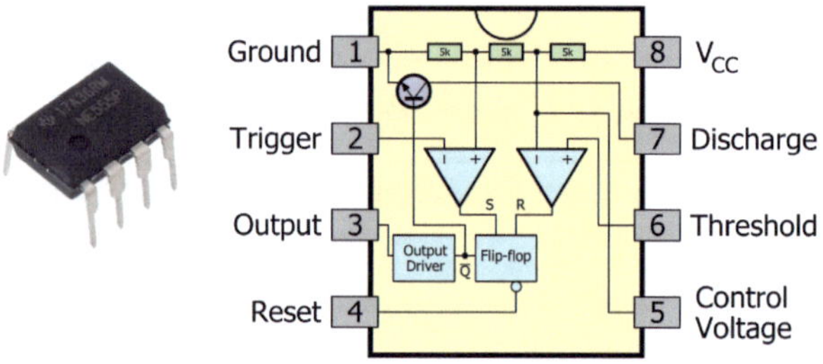

Figure 5.25 The 555 chip and its internal structure.

The basic specifications of the 555 timer are:

- Operation as astable, monostable, or as bistable.
- Timing from microseconds to hours.
- Adjustable duty cycle.
- Output can source 200 mA of current.
- Typical 3 mA current consumption.
- Operating voltage 4.5 V to 16 V.
- Output voltage V_{CC} – 1.5 V.
- High stability.

In what follows, we will develop a few projects using the 555.

5.7.1 Project 5 – Blinking an LED With the 555

Aim: This project shows how to create an astable multivibrator or free-running oscillator to flash an LED at a rate of 1 Hz.

Required components:

- 1 kΩ resistor
- 82 kΩ resistor
- 680 kΩ resistor
- 10 nF capacitor
- 1 µF capacitor
- 5 mm red LED
- 555 timer IC
- Battery cable
- Breadboard

- Jumper wires
- 9V battery

Circuit diagram: Figure 5.26 shows the schematic of the project. The timing equations for the 555 are as follows:

ON time, t1 = 0.693 × (R1 + R2) × C1
OFF time, t2 = 0.693 × R2 × C1

$$frequency = \frac{1.44}{(R1+2\cdot R2)\cdot C1} Hz$$

$$duty\ cycle = \frac{t1}{t1+t2} = \frac{R1+R2}{R1+2\cdot R2} \%$$

Where the frequency is in hertz, R1 and R2 in ohms, and capacitor C1 in farads.

In this project, the required frequency is 1 Hz. rearranging the equation gives:

$$R1 = \frac{1.44}{C1} - 2\cdot R2$$

If we use a 1 µF capacitor for C1, and choose R2 = 680 kΩ, then

$$R1 = \frac{1.44}{1\cdot 10^{-6}} - 2\cdot 680\cdot 10^{3}$$

Which gives, R1 = 80 kΩ. We will use a 82 kΩ standard-value resistor as the nearest value. Therefore, with R1 = 82 kΩ, R2 = 680 kΩ, and C = 1 µF, the frequency and duty cycle are:

$$f = \frac{1.44}{(82+2\cdot 680)\cdot 10^{3}\cdot 1\cdot 10^{-6}} = 0.998\ Hz$$

$$duty\ cycle = \frac{82+680}{82+2\cdot 680} = 52.8\,\%$$

The duty cycle is nearly symmetric (i.e. 50%).

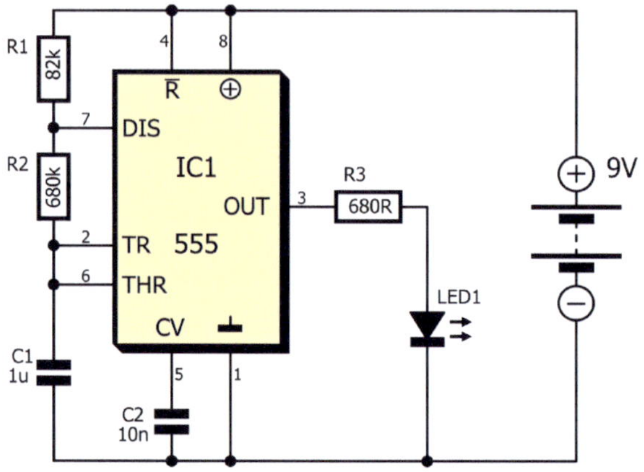

Figure 5.26 Circuit diagram of the 555 LED blinker.

Construction: Construct the project by placing the components on the breadboard as shown in Figure 5.27.

Figure 5.27 Project built on the breadboard

Procedure:

- Construct the circuit
- Connect the battery to the circuit
- The LED should flash with a frequency of about 1 Hz.

Hint: You can replace the LED with a relay and control electrical appliances easily (always be very careful when manipulating equipment connected to the mains!)

5.7.2 Project 6 – Alternating LEDs

Aim: This project is similar to the previous one but here two LEDs are used, and they blink alternately. The frequency and duty cycle are same as in the previous project.

Required components:

- 2× 1 kΩ resistor
- 82 kΩ resistor
- 680 kΩ resistor
- 10 nF capacitor
- 1 µF capacitor
- 2× 5 mm red LED
- 555 timer IC
- Battery cable
- Breadboard
- Jumper wires
- 9 V battery

Circuit diagram: Figure 5.28 shows the schematic of the project. When the output is HIGH, LED1 is forward biased, and it turns ON. LED2 blocks and therefore remains OFF. When the output is LOW, LED1 blocks. LED2 is now forward biased, and it turns ON. As a result, the two LEDs turn ON and OFF alternately.

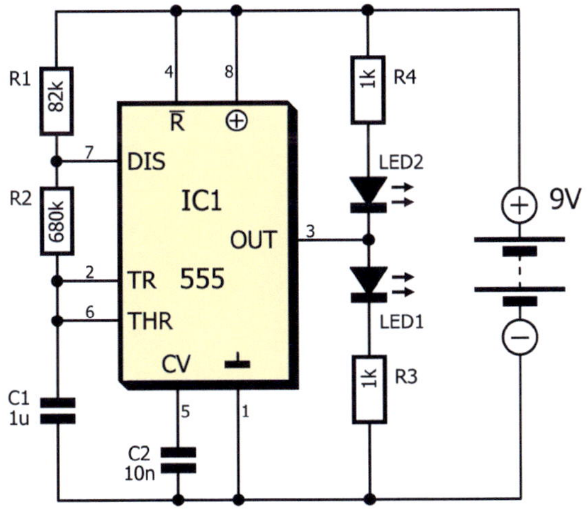

Figure 5.28 Circuit diagram of the alternating LED blinker.

Construction: Assemble the project by placing the components on the breadboard as shown in Figure 5.29.

Figure 5.29 The alternating LED blinker built on a breadboard.

Procedure:

- Build the circuit.
- Connect the battery to the circuit.
- The two LEDs should flash alternately with a frequency of about 1 Hz each.

5.7.3 Project 7 – 1 kHz Buzzer

Aim: In this project a passive buzzer is used to produce a sound with a frequency of 1 kHz.

Required components:

- 1 kΩ resistor
- 6.8 kΩ resistor
- 10 nF capacitor
- 1 µF capacitor
- 555 timer IC
- Passive buzzer
- Battery cable
- Breadboard
- Jumper wires
- 9 V battery

Circuit diagram: Figure 5.30 shows the schematic of the project. The target frequency f = 1 kHz. If we use a 0.1µF (100 nF) capacitor, then (R1 + 2 x R2) = 14.4 kΩ. Choosing R1 = 1 kΩ gives R2 = 6.7 kΩ. We choose 6.8 kΩ as the nearest standard-value resistor. The passive buzzer is connected directly to the output of the chip. Even though a passive buzzer does have polarity like e.g. an electrolytic capacitor, it is good practice to respect the polarity indicated on the device. This will also save you trouble in case you accidentally used an active buzzer. This circuit will also work with an active buzzer, but the sound will be distorted.

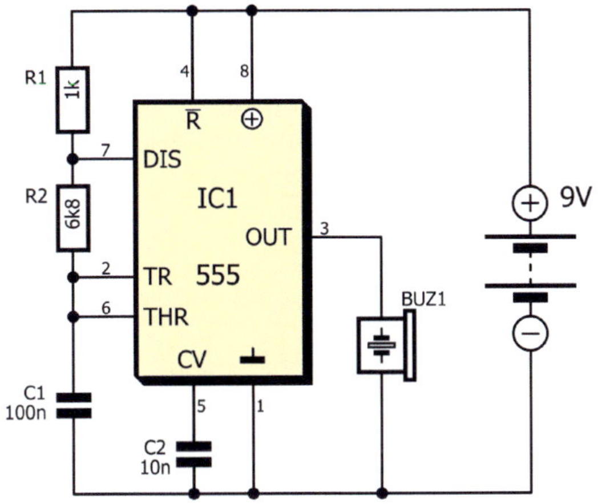

Figure 5.30 Circuit diagram of the buzzer.

Construction: Assemble the project by placing the components on the breadboard as shown in Figure 5.31.

Figure 5.31 The buzzer circuit built on a breadboard.

Procedure:

- Build the circuit.
- Connect the battery to the circuit.
- You should hear continuous tone of 1 kHz.

In this project, a passive buzzer is used as the output. We could have also used a loudspeaker, as shown in Figure 5.32. However, care must be taken not to break it with a DC voltage.

The average value of the 1 kHz output signal of the 555 is 4.5 V (50% duty cycle). This is the DC-component of the output signal and it is potentially harmful for a loudspeaker as it heats up the coil. When the coil gets too hot, it will burn, creating either a shortcut or an open circuit. Placing a capacitor in series with the loudspeaker avoids this problem as the capacitor blocks the output signal's DC component. For supply voltages higher than 9 V, select a higher resistor value (e.g. 100 Ω).

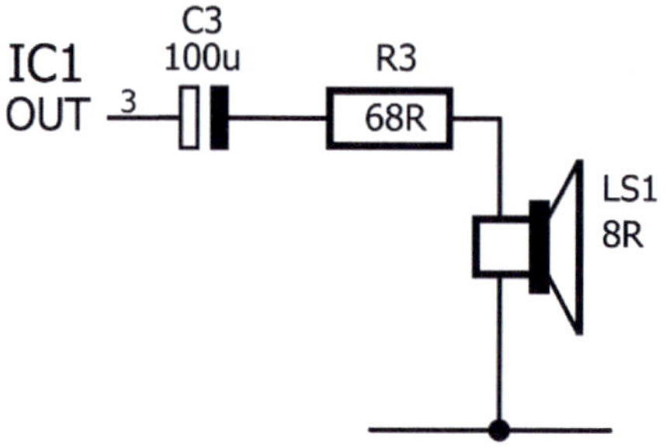

Figure 5.32 Using a loudspeaker at the output.

For higher volume and for volume control, the output circuit shown in Figure 5.33 can be used.

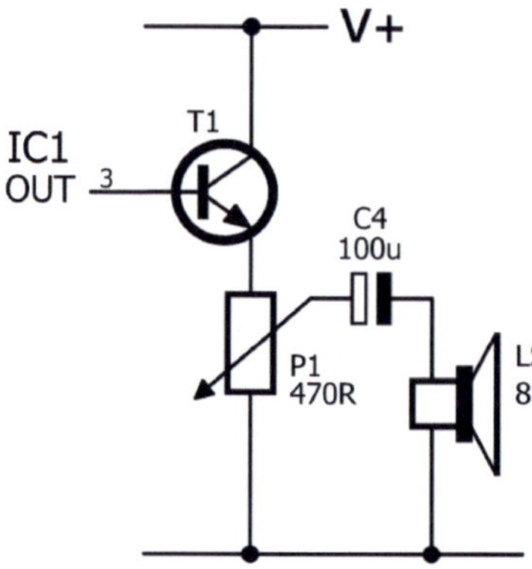

Figure 5.33 Louder output with volume control.

5.7.4 Project 8 – Continuity Tester

Aim: This is a simple continuity tester. The circuit is configured as an astable multivibrator with a frequency of 1 kHz and a buzzer, as in the previous project. The terminals of resistor R1 are used as the probe terminals to test for continuity. The buzzer sounds when there is continuity.

Required components:
The same components as in the previous project are used.

Circuit diagram: The schematic is very similar to Figure 5.30 and is shown in Figure 5.34.

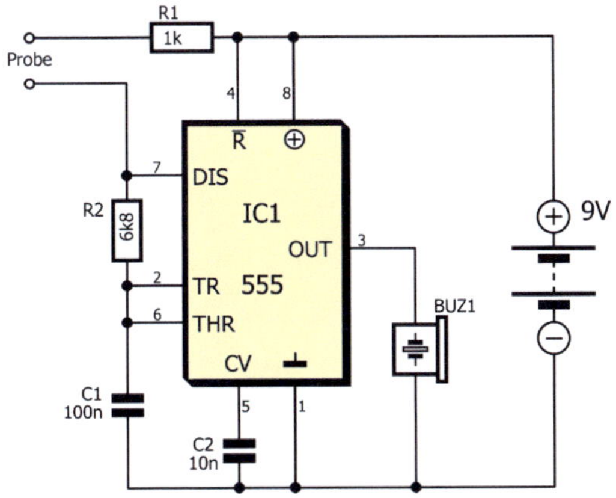

Figure 5.34 Circuit diagram of the continuity tester.
Do not use on a live circuit, i.e. a circuit that is powered on.

Procedure:

- Build the circuit.
- Connect the battery to the circuit.
- Connect the terminals of resistor R1 to the wire or trace whose continuity is to be tested. The buzzer will sound if there is continuity.

5.7.5 Project 9 – Alarm with Three Sounds

Aim: This is a basic alarm circuit with three contacts. When a contact is closed, each contact will produce a sound with a different frequency, therefore the place where the alarm occurred can be known. Contact 1, when closed, generates a 2 kHz sound, contact 2, when closed, generates a 4 kHz sound, and contact 3, when closed, generates a 6 kHz sound.

Required components:

- 3.9 kΩ resistor
- 10 kΩ resistor
- 16 kΩ resistor
- 56 kΩ resistor
- 2× 10 nF capacitor
- 555 timer IC
- 3× Push-button
- Passive buzzer
- Battery cable
- Breadboard
- Jumper wires
- 9 V battery

Circuit diagram: The circuit diagram of the project is shown in Figure 5.35. The required component values are calculated as follows:

$$f=\frac{1.44}{(R+2\cdot R2)\cdot C1}$$

Let us set C1 = 10 nF and R2 = 10 kΩ. R is then calculated for the different frequencies:

For 2 kHz

$$R=\frac{1.44}{2000\cdot 10\cdot 10^{-9}}-2\cdot 10000=52000\,\Omega$$

We can use R = 56 kΩ as the nearest standard-value resistor.

For 4 kHz

$$R=\frac{1.44}{4000\cdot 10\cdot 10^{-9}}-2\cdot 10000=16000\,\Omega$$

We can use R = 15 kΩ as the nearest standard-value resistor.

For 6 kHz

$$R=\frac{1.44}{6000\cdot 10\cdot 10^{-9}}-2\cdot 10000=4000\,\Omega$$

We can use R = 3.9 kΩ as the nearest standard-value resistor.

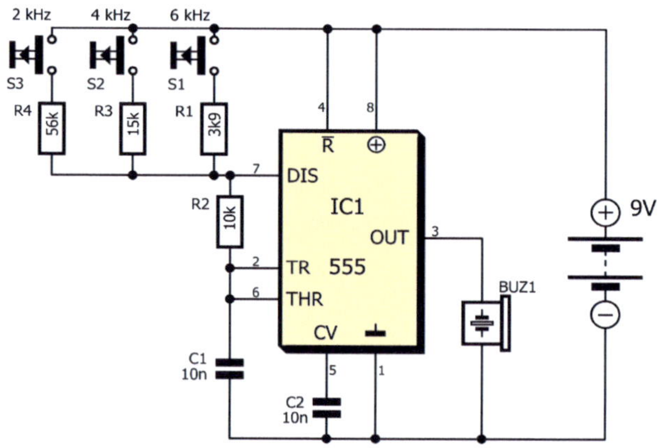

Figure 5.35 Circuit diagram of a simple alarm with three contacts.

Procedure:

- Build the circuit.
- Connect the battery to the circuit.
- Press the buttons to hear different tones.

5.7.6 Project 10 – Mosquito Repeller

Aim: It is known that mosquitos are repelled by ultrasonic frequency sound waves in the region of 40 kHz. This project generates a 40 kHz signal to drive an ultrasonic transducer.

Required components:

- 3.9 kΩ resistor
- 10 kΩ resistor
- 10 kΩ potentiometer
- 1 nF capacitor
- 10 nF capacitor
- 555 timer IC
- Battery cable
- Breadboard
- Jumper wires
- 9 V battery
- Ultrasonic transducer (not supplied)

Circuit diagram: The schematic of the project is shown in Figure 5.36. The required component values are calculated as follows:

$$f=\frac{1.44}{(R1+2\cdot R2)\cdot C1}$$

Let us set C = 1 nF. R1 and R2 are then calculated for 40 kHz as:

$$R1+2\cdot R2=\frac{1.44}{40000\cdot 1\cdot 10^{-9}}=36000$$

Therefore, if we take R1 = 3.9 kΩ, then R2 = 16.1 kΩ. We can choose e.g. 10 kΩ and then add a 10 kΩ potentiometer or trimmer in series with it to allow adjusting the frequency to exactly 40 kHz.

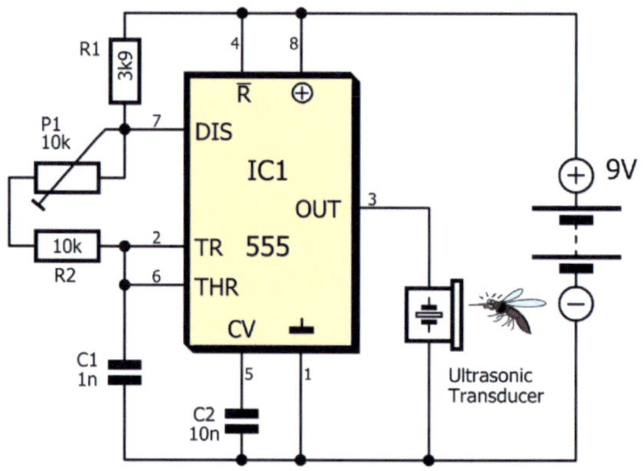

Figure 5.36 Circuit diagram of the Mosquito Repeller.

Procedure:

- Capture a mosquito.
- Build the circuit.
- Connect the battery to the circuit.
- The mosquito will make an angry face at the circuit.

5.7.7 Project 11 – Darkness Detector

Aim: In this project a light dependent resistor (LDR) is used to detect when it gets dark outside. When this happens, the buzzer starts producing a 2 kHz tone. The level of darkness that will activate the circuit is adjustable.

Required components:

- 1 kΩ resistor
- 33 kΩ resistor
- 100 kΩ potentiometer
- Light dependent resistor (LDR)
- 2× 10 nF capacitor
- 555 timer IC
- Passive buzzer
- Battery cable
- Breadboard
- Jumper wires
- 9 V battery

Circuit diagram: The schematic of the project is shown in Figure 5.37. The required component values for a 2 kHz output signal are calculated as before:

$$f = \frac{1.44}{(R1 + 2 \cdot R2) \cdot C1}$$

If we choose C = 10 nF and R1 = 3.9 kΩ, then we can calculate R2 for a frequency of 2 kHz as:

$$2 \cdot R2 = \frac{1.44}{2000 \cdot 10 \cdot 10^{-9}} - 3900 = 68100$$

Therefore, R2 = 34 kΩ. We will take 33 kΩ as the nearest standard value.

An LDR has very high resistance when it is dark and a low resistance when light falls on it. Figure 5.37 shows a typical LDR and its resistance-versus-light curve. Potentiometer P1 and the LDR form a voltage divider. When light shines on the LDR, its resistance will be low and the voltage on pin 4 of the 555 will be low too. This puts the 555 in reset mode, preventing it from producing an output signal. When it starts getting dark, the resistance of the LDR increases. This will make the voltage on pin 4 of the 555 go up (see Section 2.2.7). When this voltage rises above a certain level, the 555 will move out of reset mode and the buzzer starts to beep. The light intensity that will trigger the detector is adjustable with P1.

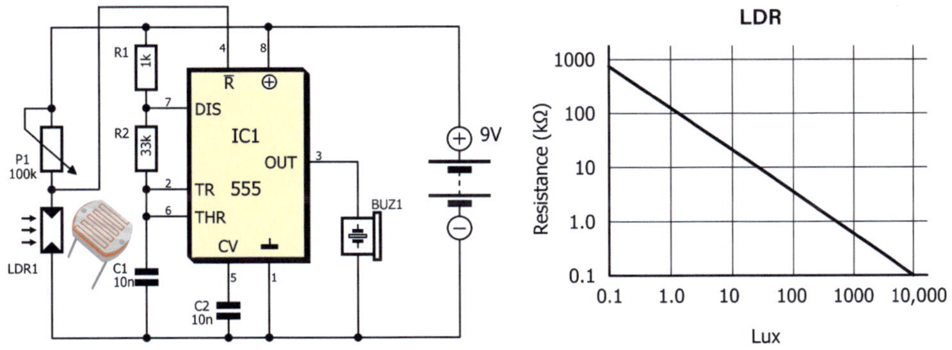

Figure 5.37 Circuit diagram of the darkness detector.
On the right, the resistance-versus-light curve of a typical LDR.

Procedure:

- Build the circuit.
- Connect the battery to the circuit.
- The buzzer should be silent when light falls on the LDR. If it isn't, turn P1 to switch it OFF. Cover the face of the LDR to make the buzzer sound.

5.7.8 Project 12 – Metronome

Aim: A metronome is a device used in music to help musicians maintain a consistent tempo, or speed. It produces regular, steady beats at a specific tempo. Beats are counted per minute (beats per minute or BPM) and ranges from less than 30 (very slow) to higher than 240 (very fast). We can convert a value in BPM to hertz by dividing it by 60. For example, 60 BPM corresponds to one beat every second, which is the same as 1 Hz. Table 5.3 shows some commonly used metronome beat rates and their musical names.

Musical name	Rate	Beats per minute	Frequency (Hz)
Largo	Very slow	40	0.66
Larghetto	Rather slow	60	1.0
Andante	Slow	90	1.5
Moderato	Medium	120	2.0
Allegro	Fast	150	2.5
Presto	Very fast	180	3.0
Prestisimo	Extremely fast	208	3.46

Table 5.3 Metronome beats and equivalent frequency in Hz.

In orde to obtain the required frequencies for our metronome, we will keep the timing capacitor C1 and resistor R1 fixed and change resistor R2 to generate frequencies in the range 0.66 Hz to 3.46 Hz. Let us set C = 3.3 μF and R1 = 22 kΩ. The required values of R2 are then calculated as:

$$R2 = \frac{1}{2}\left(\frac{1.44}{f \cdot C1} - R1\right)$$

For 0.66 Hz

$$R2 = \frac{1}{2}\left(\frac{1.44}{0.66 \cdot 10 \cdot 10^{-6}} - 22000\right) = 98091\,\Omega$$

In other words, R2 = 98 kΩ.

For 3.46 Hz

$$R2 = \frac{1}{2}\left(\frac{1.44}{3.46 \cdot 10 \cdot 10^{-6}} - 22000\right) = 9809\,\Omega$$

And so R2 = 9.8 kΩ.

Thus, if we use a 100 kΩ potentiometer for R2, we can obtain all the required metronome beats.

Required components:

- 10 kΩ resistor
- 100 kΩ potentiometer
- 10 nF capacitor
- 10 μF capacitor
- 555 timer IC
- Passive buzzer
- Battery cable
- Breadboard
- Jumper wires
- 9 V battery

Circuit diagram: The circuit diagram of the project is shown in Figure 5.38. In this circuit a passive buzzer is used, but it can be replaced with a loudspeaker and its series capacitor as before.

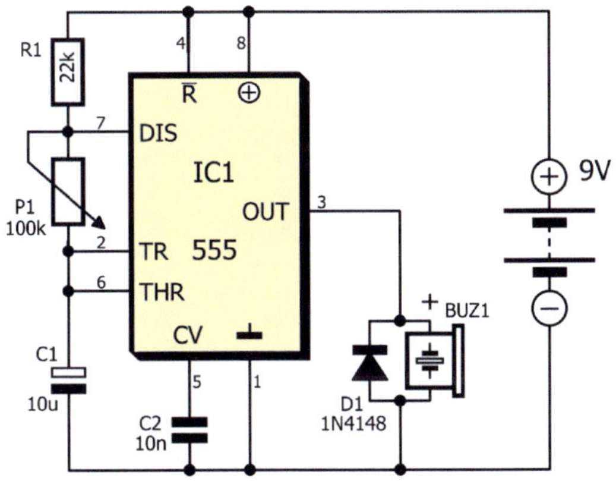

Figure 5.38 Circuit diagram of the metronome.

Procedure:

- Build the circuit.
- Connect the battery to the circuit.
- Change the potentiometer setting to listen to different beat rates.

Hint: You can make a circular dial with the beat rates marked on it. Use a mechanical metronome to accurately label the markings on the dial. Alternatively, there are many metronome apps for smartphones and on the internet, and you can use one of these to label the markings on your dial.

5.7.9 Project 13 – LED Timer

Aim: In this project, we will build a monostable multivibrator. When the button is pressed, the LED is turned ON for 5 seconds. The output of the monostable goes HIGH for the specified time and then goes LOW.

Required components:

- 1 kΩ resistor
- 10 kΩ resistor
- 47 kΩ resistor
- 10 nF capacitor
- 100 μF capacitor
- 5 mm red LED
- 555 timer IC
- Battery cable
- Breadboard
- Jumper wires
- 9 V battery

Circuit diagram: The circuit diagram of the project is shown in Figure 5.39. The duration of the output pulse of a 555 monostable multivibrator is calculated as follows:

$$T = 1.1 \times R1 \times C1$$

Where T is in seconds, R1 is in ohms, and C1 is in farads. For a pulse duration of 5 seconds while choosing C1 = 100 μF, the resistor value is:

$$R1 = 5 / (1.1 \times 100 \times 10^{-6}) = 45.5\ k\Omega.$$

The nearest standard value is 47 kΩ, which gives 5.17 seconds.

When the push-button is not being pressed the voltage on pin 2 of the 555 is 9 V (assuming there is no or very little current flowing into pin 2) and the 555 is idle. When the button is pressed, the voltage on pin 2 drops to zero, which triggers the monostable multivibrator. Now the output goes HIGH and the LED lights up. The value of R2 is not very important and can be chosen as 10 kΩ. It limits both the current through the push-button and into pin 2.

A 1 kΩ current limiting resistor is used at the output to limit the current through the LED.

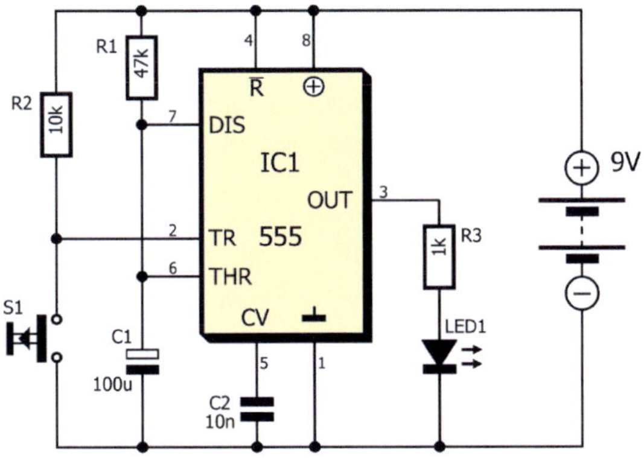

Figure 5.39 Monostable circuit.

Construction: Figure 5.40 shows the project built on a breadboard.

Figure 5.40 Project built on a breadboard.

Procedure:

- Build the circuit and connect the battery.
- Press the button. The LED should turn ON for 5 seconds and should then turn OFF until the button is pressed again.

5.7.10 Project 14 – Outside Light with Long Delay

Description: In this project a monostable multivibrator controls a relay that is connected to a lamp. The lamp has its own AC or DC power supply. The time delay is set to 60 seconds so that pressing the button will turn ON the lamp for 60 seconds.

Required components:

- 10 kΩ resistor
- 120 kΩ resistor
- 10 nF capacitor
- 470 µF capacitor
- 1N4148 diode
- 555 timer IC
- 9 V SPDT relay
- Battery cable
- Breadboard
- Jumper wires
- 9 V battery
- Lamp (not supplied)

Circuit diagram: The circuit diagram of the project is shown in Figure 5.41. The timing of a 555 monostable is calculated as follows:

$$T = 1.1 \times R1 \times C1$$

Where T is in seconds, R1 is in ohms, and C1 is in farads. For a pulse duration of 60 seconds while choosing C1 = 470 µF, the resistor value is:

$$R1 = 60 / (1.1 \times 470 \times 10^{-6}) = 116 \text{ k}\Omega.$$

The nearest standard value is 120 kΩ, which gives a pulse duration of 62 seconds.

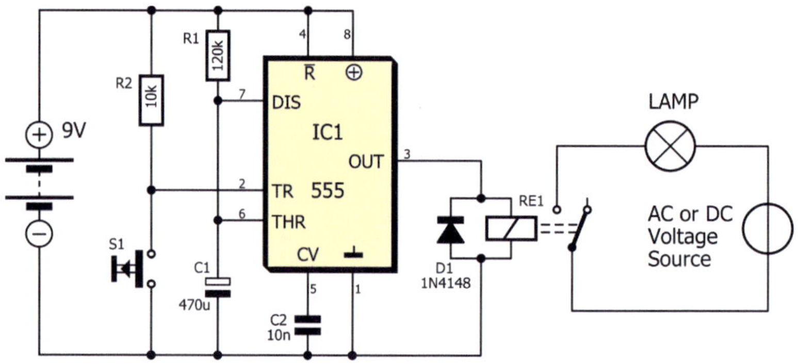

Figure 5.41 Circuit diagram of the outside light timer.

A diode is used in parallel with the relay in reverse biased mode to protect the output circuit of the 555 when the relay switches OFF (see also section 3.7).

Warning! Ask a qualified electrician for assistance if you want to connect the relay contacts to the mains supply. Do not electrocute yourself!

Procedure:

- Build the circuit and connect the battery.
- Press the button. The relay (and the lamp) should turn ON for about 60 seconds before turning OFF.

Hint: You could use other loads instead of a lamp at the relay outputs, e.g. a motor or a fan.

5.7.11 Project 15 – Improved Metronome

Description: In this project we combine the monostable multivibrator from the previous two projects with Project 12, the metronome (Section 1.5.8). The inconvenience of the metronome described in that project is that the duration of its click sound depends on the beat rate. By adding a monostable multivibrator to the metronome's output, we can obtain a click sound with a constant duration, independent of the beat rate.

Required components:

- 2× 10 kΩ resistor
- 100 kΩ potentiometer
- 2× 10 nF capacitor
- 2× 10 µF capacitor
- 2× 555 timer IC
- Passive buzzer
- Battery cable
- Breadboard
- Jumper wires
- 9 V battery

Circuit diagram: The circuit diagram of the project is shown in Figure 5.42 The timing of the 555 monostable is calculated as follows:

$$T = 1.1 \times R2 \times C3$$

Where T is in seconds, R2 is in ohms, and C3 is in farads. The fastest click rate was set to 3.46 Hz. This means that a click sound may not be longer than 289 ms, otherwise it is impossible to distinguish clicks as it would be on continuously. If we set it to about 100 ms, then we are safe. Therefore, if we choose C3 = 10 µF, then the value of R2 is:

$$R2 = 0.1 / (1.1 \times 100\times10^{-6}) = 9.1\ k\Omega.$$

A good standard value is 10 kΩ, which gives a pulse duration of 110 ms.

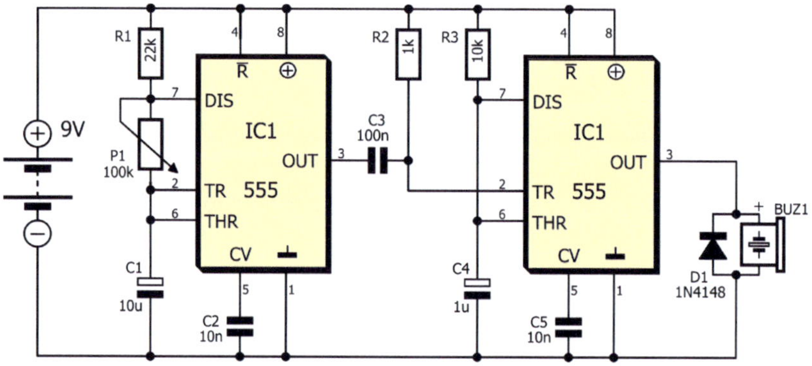

Figure 5.42 Circuit diagram of the improved metronome.

Procedure:

- Build the circuit.
- Connect the battery.
- Change the potentiometer setting to listen to different beat rates.

Figure 5.43 The improved metronome built on a breadboard.

5.7.13 The Bistable 555

The 555 timer can also be operated in bistable mode, i.e. as a flip-flop. Figure 5.44 shows the circuit diagram of a basic bistable multivibrator. The circuit is very simple with only two resistors and no capacitors. Here, the bistable has two trigger inputs: Set and Reset. A pulse on the Set input sets the output to the HIGH state. The output remains in this state until the Reset input is triggered. Then the output returns to the LOW state. Notice that both the Set and Reset inputs of the 555 are active LOW, meaning that a LOW voltage triggers the action.

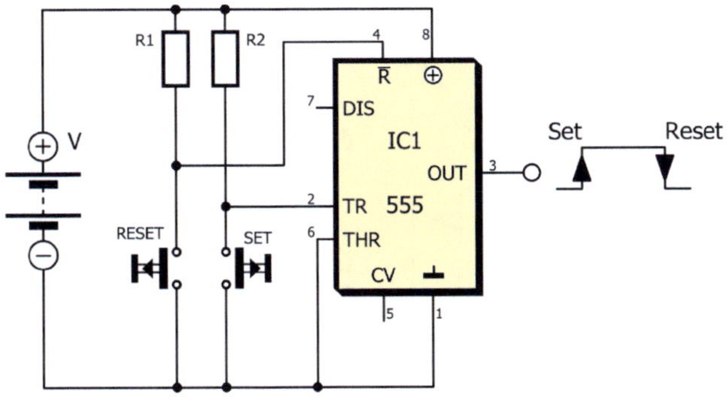

Figure 5.44 The 555 as a Set/Reset (SR) flip flop.

5.8 Project 15 – Two-Transistor Blinky

Description: This is a simple blinker project using two transistors and an LED. The blink rate is set using a potentiometer.

Required components:

- 100 Ω resistor
- 2× 1 kΩ resistor
- 10 kΩ potentiometer
- 2× BC337 transistor
- 100 µF capacitor
- LED
- Battery cable
- Breadboard
- Jumper wires
- 9 V battery

Circuit diagram: The schematic of the project is shown in Figure 5.45. The circuit is based on a relaxation oscillator. A relaxation oscillator is a non-linear oscillator than can generate a non-sinusoidal repetitive output waveform.

Capacitor C1 charges through the potentiometer and resistor R1. When the voltage over C1 reaches a threshold value, transistor T1 starts to conduct. This turns on T2 and the LED switches ON. When T1 conducts, C1 starts to discharge through R2. When the voltage over C1 reaches a lower threshold value, transistor T1 stops conducting. T2 turns off, the LED switches OFF and the process starts all over again.

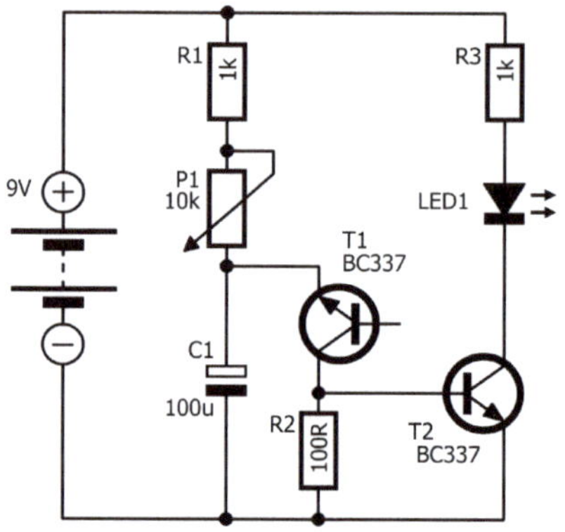

Figure 5.45 Circuit diagram of a simple relaxation oscillator.

Procedure:

- Assemble the circuit and connect the battery.
- You should see the LED blink periodically. Turn the potentiometer to adjust the blink rate.

5.9 Fixed-Output Voltage Regulator ICs

Many electronic circuits and microcontroller development boards (e.g. Arduino UNO) operate from +5 V power supplies. Therefore, it may be necessary to obtain +5 V from a 9 V battery or from another DC source. As their name implies, fixed-output voltage regulator ICs are used to provide fixed voltages in electronic circuits. The input voltages of these ICs are higher than their output voltages. The most popular IC families for fixed output voltages are the 78xx and 79xx series where 'xx' indicates the output voltage. As an example:

- 7805: Fixed +5 V output voltage
- 7812: Fixed +12 V output voltage
- 7905: Fixed -5 V output voltage
- 7912: Fixed -12 V output voltage

78xx and 79xx ICs can provide up to 1.0 A of current. There are also smaller, low-power versions of these ICs. For example, the 78L05 provides +5 V, but its current is limited to 100 mA. Similarly, the 78L12 provides +12 V at 100 mA max.

The specifications of the 7805 5 V regulator IC are:

- Minimum input voltage: +7 V
- Maximum input voltage: +35 V
- Maximum current: 1 A
- Maximum output voltage: +5.2 V
- Minimum output voltage: +4.8 V

Notice that the input voltage must be at least +7 V for the 7805 to function properly. For the 7812 and 78L12 versions, the input voltage must be in the range from +14.5 V to +27 V.

Figure 5.46 shows how the 7805 can be used in a circuit to turn a 9 V battery into a +5 V power supply. The low-power versions 78Lxx and 79Lxx can be used in the same configuration as the 78xx and 79xx versions.

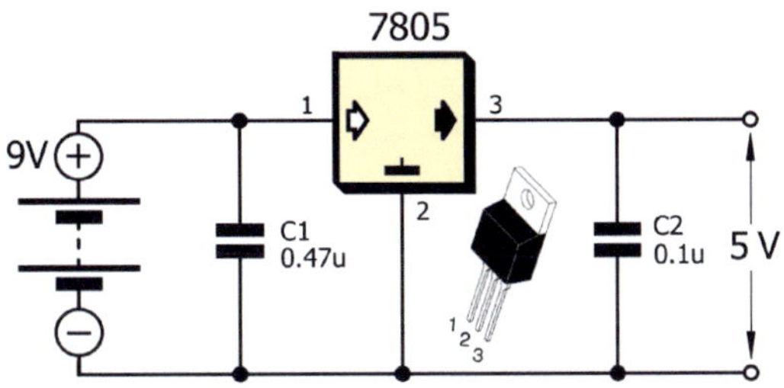

Figure 5.46 Using the 7805 fixed-output linear voltage regulator.

78xx/79xx Heating Problems

The 78xx/79xx devices can have heating problems if the input voltage is much larger than the output voltage, or if large currents are drawn. The power dissipated by the IC can be calculated as follows:

The power consumption of the IC, $P = (V_{IN} - V_{OUT}) \times I_{OUT}$

If for example, the input voltage of an 7805 is 20 V and the output current is 0.5 A, then,

$$P = (20 - 5) \times 0.5 = 7.5 \text{ W}$$

This means that the device must dissipate 7.5 W of power. A heatsink will be required to keep the device at an acceptable temperature, otherwise it will break. Notice that the power consumed by the load is only 5 × 0.5 = 2.5 W. Therefore, 5 W is wasted as heat!

5.10 Reverse Polarity Protection

Connecting a power supply with incorrect polarity is a common and easy mistake to make. This can easily damage electronic components unless they are protected with reverse-polarity protection circuitry. Reverse-polarity protection is usually created with a diode. In its simplest form, a diode can be connected in series with the power supply as shown in Figure 5.47. In this circuit, if the power supply is connected with the polarity reversed, then the diode cannot conduct. No current can flow, and the circuit is safe. It is important that a diode is chosen with the correct wattage so that it can handle the maximum expected load current.

The main disadvantage of the circuit is that about 0.7 V will drop across the diode. This lowers the input voltage of the circuit by 0.7 V, which can make it too low for certain applications. Sometimes a Schottky diode can be used in such cases since its voltage drop is only around 300 mV.

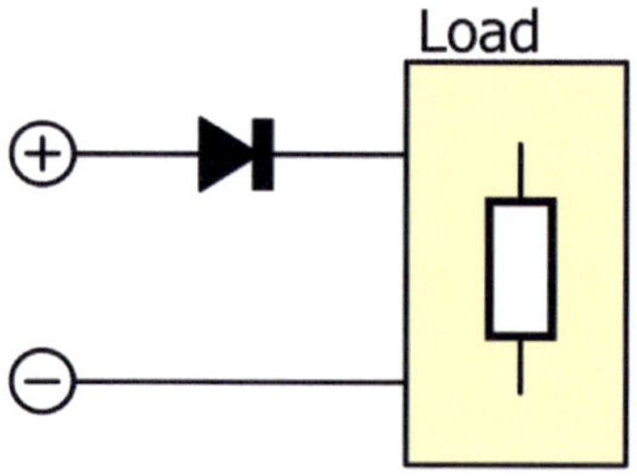

Figure 5.47 Reverse polarity protection circuit.

Chapter 6 • Operational Amplifiers

6.1 Overview

The Operational Amplifier (or op-amp) is a voltage difference amplifier designed to be used with external components to provide functions such as voltage amplification, filtering, waveform shaping, oscillator circuits and so on. Operational amplifiers are available as integrated circuits (ICs), but they can also be constructed from discrete elements like transistors, diodes, resistors, and capacitors.

In this chapter, we will be looking at the basic principles of operational amplifiers and learn how various functions can be performed using them.

6.2 Operational Amplifier Structure

An operational amplifier is a three-terminal device (excluding the power terminals) which consists of two very high impedance inputs labelled as '–' and '+'. The '–' input is called the inverting input, while the '+' input is called the noninverting input. The third terminal is the output. There are a huge number of operational amplifier ICs available, such as general purpose, precision, high-speed, low-noise, high-voltage, etc. Some ICs include only one op-amp, while others include two (the dual op-amp) or more (usually four: the quad op-amp) in a single IC.

Figure 6.1 shows a typical operational amplifier IC with its equivalent circuit.

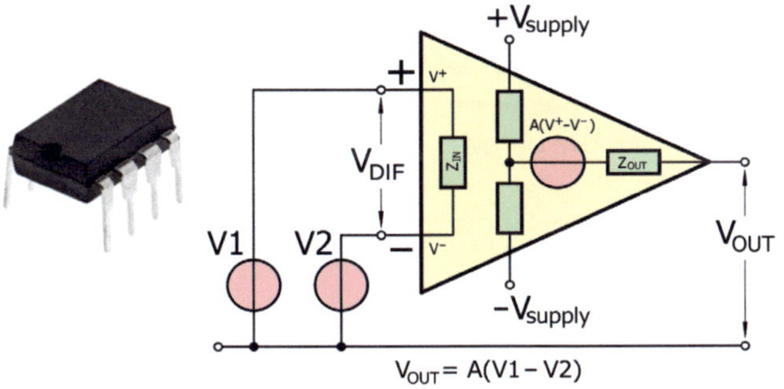

Figure 6.1 The equivalent circuit of the operational amplifier and a device in a DIP-8 package.

The op-amp amplifies the difference between the voltages on its noninverting and inverting inputs:

$$V_o = A(V_+ - V_-)$$

Here, 'A' is the open-loop gain of the op-amp. Dividing both sides by A gives:

$$\frac{V_o}{A_i} = V_+ - V_-$$

The value of A is very large (infinite for an ideal op-amp), meaning that this equation can be simplified to:

$$V_+ = V_-$$

This is an important equation to keep in mind when analysing op-amp circuits. It implies that an op-amp will adjust the voltage on its output in such a way that the voltages on its inputs become the same. The only way that this can be done is by providing a path from the output of the op-amp to at least one of its inputs. This is called **feedback**. Almost every op-amp circuit has some form of feedback.

The characteristics of an **ideal** operational amplifier are summarized here:

- Infinite open-loop gain. Real values are around 200,000 or more.

- Infinite input impedance; therefore, no current flows into its inputs. In reality, the input impedance is very high but not infinite and a tiny current will flow into the inputs.

- Zero output impedance. In reality, the output impedance is very low but not zero.

- Infinite bandwidth, i.e. it can amplify signals of any frequency. This is, of course, limited in reality.

- Zero offset voltage, i.e. the output voltage is zero if the same voltage is applied to its inverting and noninverting terminals. In reality, there will be a small offset voltage present on the output.

In the next sections, we will be looking at how operational amplifiers can be used in various electronic circuits.

6.3 Inverting Amplifier

Figure 6.2 shows an inverting operational amplifier circuit. Two resistors RF and R1 are used to set the voltage gain of the amplifier. RF is used as a negative feedback resistor. The noninverting input is connected to the common (usually called ground) and its voltage V_+ can be considered to be 0 V.

As explained above, since the circuit provides feedback through RF, the voltages on the op-amp's inputs are equal. Therefore, since $V_+ = 0$ V, V_- is also equal to 0 V.

Supposing that no current flows into the inputs of the op-amp, we can write:

$$\frac{V_i}{R1} = \frac{-V_o}{RF}$$

The minus sign is because V_- is 0 V. The current flows from V_i through R1 to zero, and then from zero through RF to V_O. Therefore, the current through RF is opposite to the current through R1, which is represented by the minus sign.

The voltage gain A_V of the circuit is found by rearranging the previous equation:

$$A_V = \frac{V_o}{V_i} = \frac{-RF}{R1}$$

Do not confuse A_V with the open-loop gain A.

The output voltage is given by:

$$V_o = \frac{-RF}{R1} V_i$$

The negative sign in the equation indicates that the output voltage is inverted (i.e. 180° out of phase) compared to the input voltage.

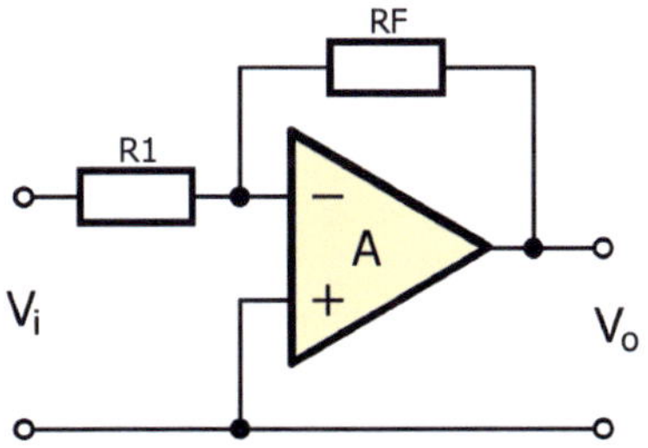

Figure 6.2 Inverting operational amplifier (power supply connections not shown)

Example
We want to amplify a 1 mV signal to −100 mV using an inverting operational amplifier. Calculate the component values and draw the required circuit.

Solution
The required voltage gain $A_V = -100/1 = -100$. If we choose R1 = 1 kΩ, then RF = 100 kΩ. The required circuit is shown in Figure 6.3.

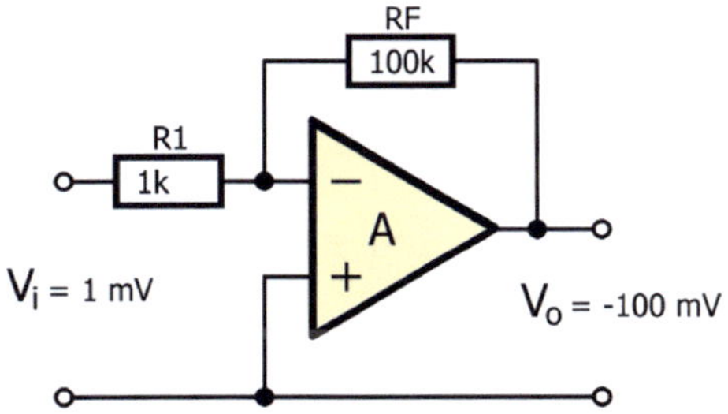

Figure 6.3 Operational amplifier with a voltage gain of −100.

6.4 Noninverting Amplifier

Figure 6.4 shows a noninverting operational amplifier circuit. Again, two resistors RF and R1 are used to set the voltage gain of the amplifier.

As for the inverting amplifier, the circuit provides feedback through RF and therefore the voltages on the op-amp's inputs are equal. This time, V_- is not connected to the common, and it is therefore not equal to 0 V. But it is equal to V_+, which is V_i.

Supposing that no current flows into the inputs of the op-amp, we can write:

$$\frac{V_i}{R1}=\frac{V_o-V_i}{RF}$$

The voltage gain AV of the circuit is found by rearranging the previous equation:

$$A_V=\frac{V_o}{V_i}=1+\frac{RF}{R1}$$

The output voltage is given by:

$$V_o=\left(1+\frac{RF}{R1}\right)V_i$$

Note that the output voltage is in phase with the input voltage.

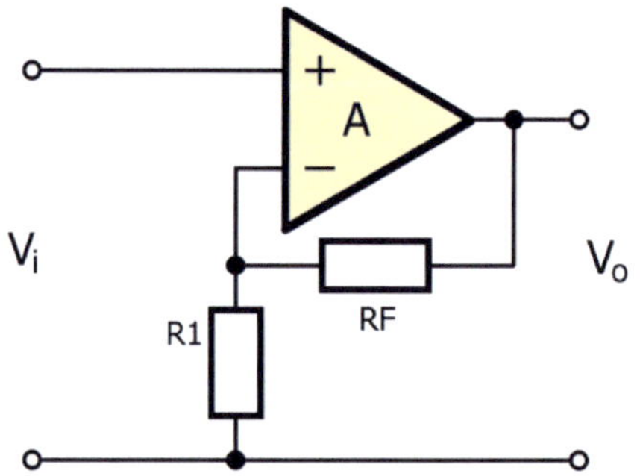

Figure 6.4 Inverting operational amplifier (power supply connections not shown).

Example

Amplify a 1 mV signal to 100 mV using a noninverting operational amplifier. Calculate the component values and draw the required circuit.

Solution

The required voltage gain A_V = 100/1 = 100. If we choose R1 = 1.8 kΩ, then,

100 = 1 + RF / 1.8

Which gives, RF = 178.2 kΩ. As we can accept an error of 1%, we choose RF = 180 kΩ as the nearest standard value. Figure 6.5 shows the required circuit.

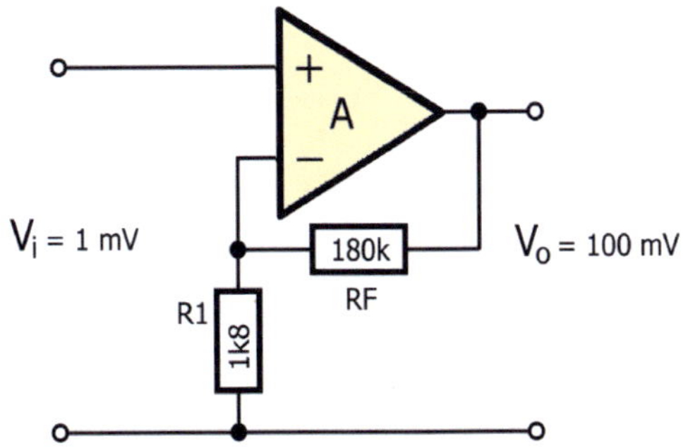

Figure 6.5 Noninverting amplifier with a voltage gain of 100.

6.5 Voltage Follower

A voltage follower has a voltage gain of unity ($A_V = 1$). The circuit provides a high input impedance with a very low output impedance; it is an impedance adapter. This op-amp configuration is a buffer that turns a high output impedance into a low impedance. It creates a (strong) copy of a signal without influencing the (weak) signal.

Figure 6.6 shows the circuit of a voltage follower. No resistors are used in the circuit. As the output is connected to the inverting input, there is feedback. Hence, V_- is equal to V_+, which in turn is equal to V_i. Therefore, $V_o = V_i$.

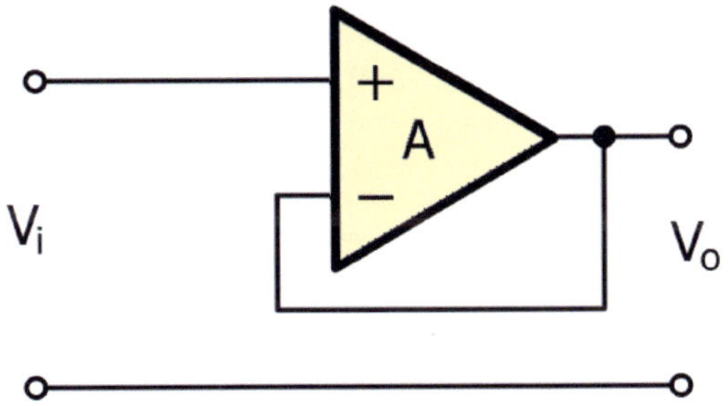

Figure 6.6 Voltage follower circuit.

6.6 Summing Amplifier

Figure 6.7 shows a summing amplifier. In this circuit, the output voltage is equal to the negative sum of the input voltages and is given by (only three inputs are shown here, but there can be more or fewer):

$$V_o = -\left[\frac{RF}{R1}V1 + \frac{RF}{R2}V2 + \frac{RF}{R3}V3\right]$$

If R1 = R2 = R3 = R, then the output voltage equation becomes:

$$V_o = -\frac{RF}{R}[V1 + V2 + V3]$$

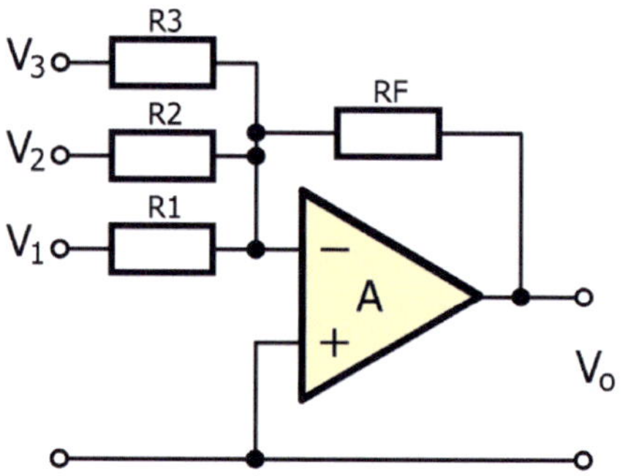

Figure 6.7 Summing amplifier circuit.

If we further let RF = R, then the output voltage becomes:

$$Vo = -[V1 + V2 + V3]$$

Note the minus sign. This summing amplifier is an inverting amplifier.

As an exercise, apply the theory presented at the beginning of this chapter to derive the three equations above.

6.7 Difference Amplifier

A difference amplifier is basically a voltage subtractor, where the output voltage is equal to the difference of the two input voltages. Figure 6.8 shows the circuit diagram of a difference amplifier.

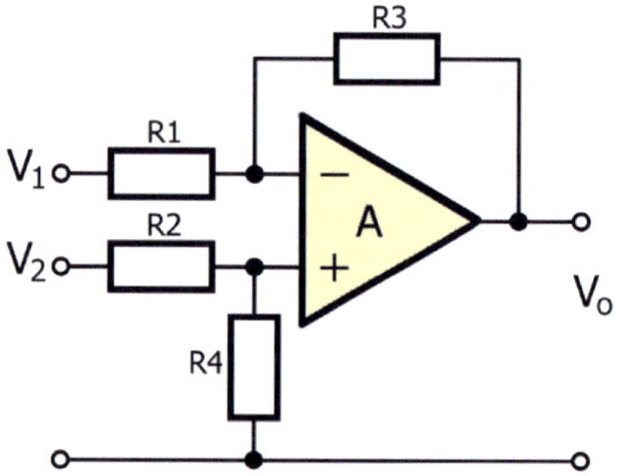

Figure 6.8 Difference amplifier.

If we let R1 = R2 and R3 = R4, then the output voltage is given by:

$$Vo = \frac{R3}{R1}(V2 - V1)$$

If further R1 = R3, then the output voltage is the difference of the two input voltages:

$$Vo = V2 - V1$$

As an exercise, apply the theory presented at the beginning of this chapter to derive the two equations above.

6.8 Integrator

In an integrating amplifier, the output voltage is proportional to the integral of the input voltage over time. A capacitor is used in the feedback path as shown in Figure 6.9.

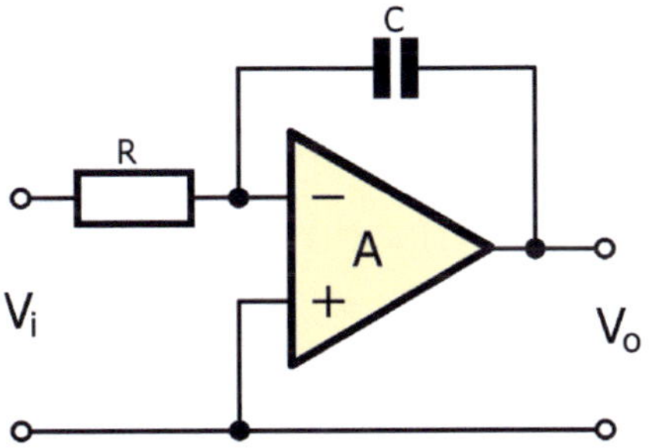

Figure 6.9 The integrator.

The integrator is a variation on the inverting amplifier. To calculate the output voltage V_o, we can use the same equations, but with the impedance of RF replaced by the impedance of C, Z_C:

$$\frac{V_i}{R} = \frac{-V_o}{Z_C}$$

Furthermore (see Chapter 2),

$$\frac{V_o(t)}{Z_C} = I_C(t) = C\frac{dV_o(t)}{dt}$$

Combining these two equations gives:

$$\frac{V_i(t)}{R} = -C\frac{dV_o(t)}{dt}$$

Which can be rearranged to obtain the output voltage:

$$V_o(t) = \frac{-1}{RC}\int V_i(t)\,dt$$

6.9 Differentiator

In a differentiator amplifier, the output voltage is proportional to the differential of the input voltage over time. A capacitor is used in the input path as shown in Figure 6.10.

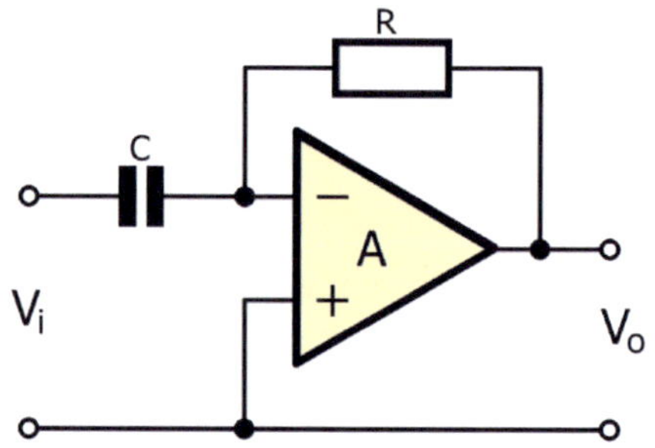

Figure 6.10 The differentiator.

Similar to the integrator, the output voltage is now given by:

$$V_o(t) = -RC\frac{dV_i(t)}{dt}$$

Again, note the minus sign. This differentiator too is a variation on the inverting amplifier.

6.10 Comparator

A comparator compares the voltages on its inputs and indicates which one is higher than the other. Figure 6.11 shows an operational amplifier wired as a comparator. Here, the output will go HIGH when $V_i > V_r$. When $V_i < V_r$ the output will go LOW. V_r is a reference voltage created with voltage divider R1 and R2:

$$V_r = \frac{R2}{R1+R2}V$$

So, what happens when $V_i = V_r$? In an ideal world, the output would settle halfway between LOW and HIGH. In the real world, however, to a comparator, V_i is never equal to V_r even if they come from the same output. This is due to tiny differences and tolerances in the inputs. When $V_i = V_r$, one of the two will beat the other in size by a tiny bit, and the comparator's output will reflect this. Which one, we don't know in advance, but the output will be either LOW or HIGH.

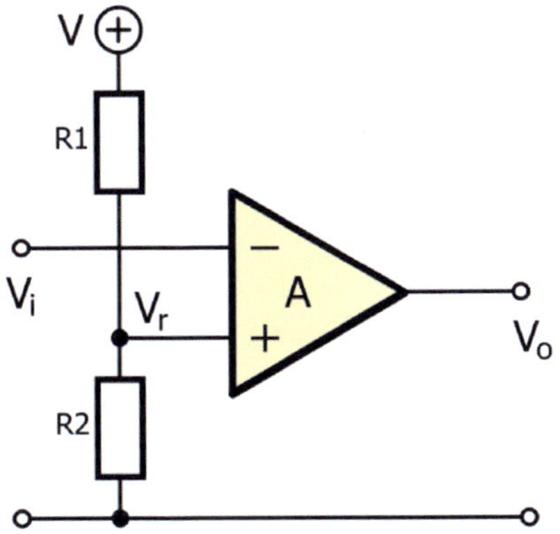

Figure 6.11 Comparator circuit.

6.11 Summing Amplifier Simulation

In this simulation, we will be adding three voltages (1 V, 2 V and 4 V) at the input and expect to get 7 V at the output. The steps are:

- Draw the circuit as shown in Figure 6.12. Set the input Voltage Sources to 1 V, 2 V and 4 V as shown in the figure.
- Set all the resistors to 1 kΩ so that the output voltage is simply the sum of the input voltages.
- Set the two battery voltages to 12 V. Set the operational amplifier type to uA741.
- Notice that the common GND is the point in the middle of the two batteries.
- Click **Analysis → DC Analysis → Table of DC results**
- You should see the voltages at the nodes as shown in Figure 6.13. The voltage at the output of the operational amplifier (VP_7) is −7 V, as expected.

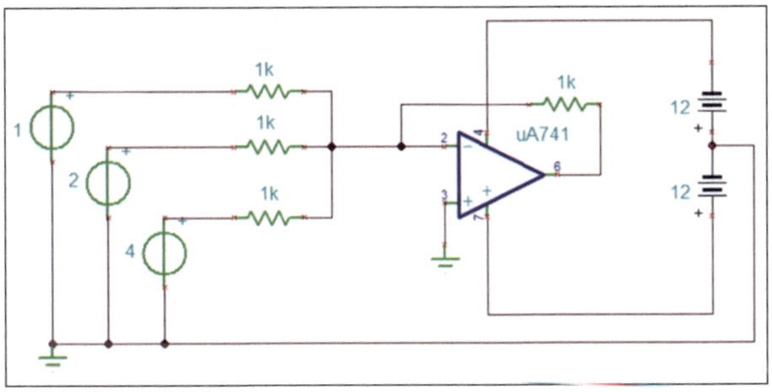

Figure 6.12 Simulation circuit.

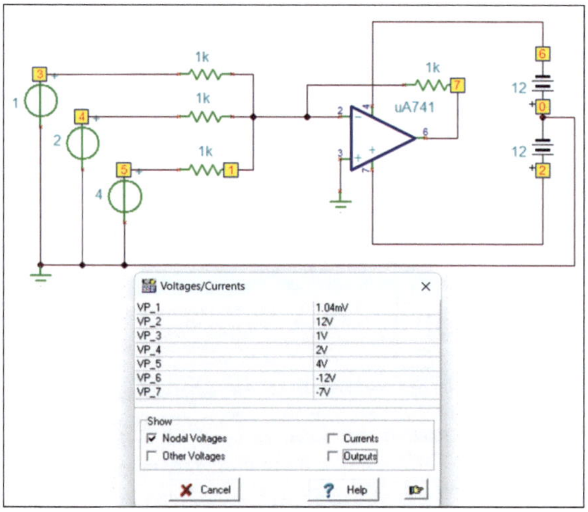

Figure 6.13 DC voltages in the circuit

6.12 Project 1 – Comparator With LED

Aim: The aim of this project is to show how a comparator functions in real-life. In this circuit, a fixed voltage is applied to the positive input of a comparator IC. A potentiometer is connected to the negative input. The output of the comparator is connected to an LED through a current limiting resistor (R3). Normally, the output is at 9 V and the LED is OFF. Turning the potentiometer increases the voltage at the negative input. When it becomes higher than the voltage at the positive input, the output of the comparator goes to 0 V and the LED turns ON.

Required components:

- 3× 1 kΩ resistor
- 10 kΩ potentiometer
- 5 mm red LED
- LM393 comparator IC
- Battery cable
- Breadboard
- Jumper wires
- 9 V battery

The LM393

This is a dual comparator IC. Its pin configuration is shown in Figure 6.14. The specifications of this comparator are:

- 2 V to 36 V operating voltage
- Output sinking current 16 mA
- Output saturation voltage 150 mV
- Response time 1.3 μs

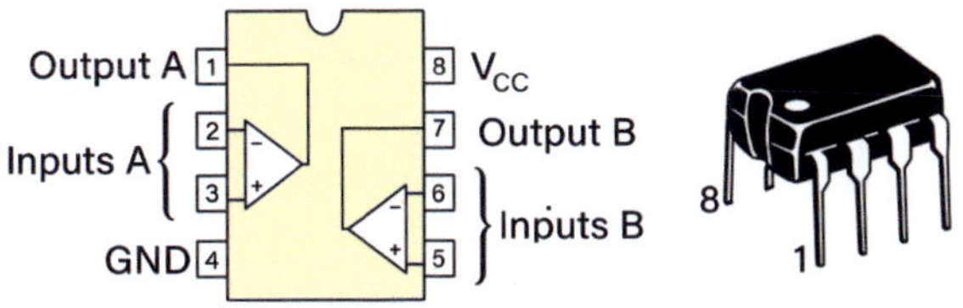

Figure 6.14 The LM393 is a dual comparator.

Circuit diagram: Figure 6.15 shows the circuit diagram of the project.

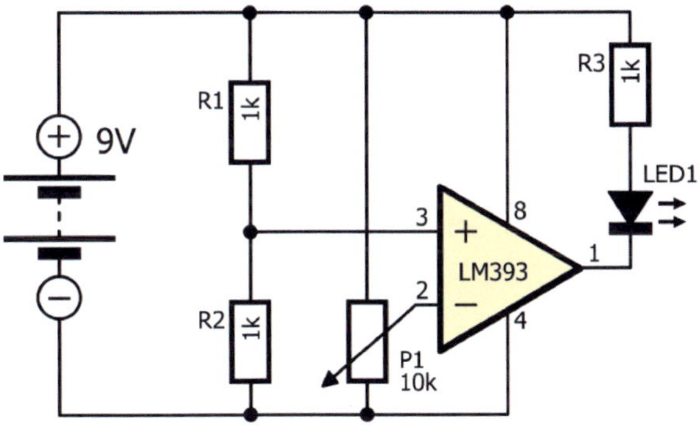

Figure 6.15 A simple voltage comparator circuit.

The DC voltage at the positive input of the comparator is 4.5 V since the two resistors R1 and R2 have the same value. If the voltage at the wiper of the potentiometer is 0 V then the output of the comparator is 9 V and the LED is OFF. Turning the potentiometer increases the voltage at its wiper. When it passes 4.5 V the output of the comparator changes to 0 V, which turns ON the LED in current sinking mode.

Construction: Build the project by placing the components on the breadboard as shown in Figure 6.16.

Figure 6.16 Assemble the project on a breadboard.

Procedure:

- Place the components on the breadboard as shown in Figure 6.15 and connect the battery.
- If the LED is OFF, turn the potentiometer until the LED turns ON. If, on the other hand, the LED is ON, turn the potentiometer until it turns OFF. At this point, the voltage at the wiper of the potentiometer will be close to 4.5 V. Turn the potentiometer to the other side until the LED changes state again.

Chapter 7 • Logic Gates

7.1 Overview

Logic gates are the basic building blocks of digital systems. They are available as ICs with one or several logic gates in each IC. Logic gates are named after their function: AND, OR, NOT, NAND, NOR, etc. In this chapter, we will look at the principles of various logic gates and learn how they can be used in simple projects.

Logic gates perform Boolean algebra, a branch of algebra. Values in Boolean algebra can take on only one of two values: FALSE or TRUE. In digital electronics, these values are represented by '0' and '1'. Usually, TRUE or '1' refers to a HIGH voltage (e.g. 3.3 V, 5 V, etc.) while FALSE or '0' refers to a LOW voltage (e.g. close to 0 V), but the inverse can be true too (called 'inverted logic').

The functions performed by logic gates are often represented by a table, known as the Truth Table. Such a table shows the relationship between the function's inputs and outputs.

7.2 Commonly Used Logic Gates

Logic gates can have several inputs and outputs. In this section, we will assume that the gates have two inputs and one output (except the NOT gate that has only one input).

Like resistors and capacitors, there are two ways to draw logic gates: as American ANSI or as European IEC symbols. You can use whichever you prefer but avoid mixing them in a schematic.

7.2.1 AND Gate

In Boolean algebra, the AND function is a multiplication: FALSE × FALSE = FALSE; FALSE × TRUE = FALSE; TRUE × TRUE = TRUE. Therefore, the output of the AND gate is only TRUE when both (or all) of its inputs are TRUE. The output is FALSE in all other situations. The symbol and the truth table of a two-input AND gate are shown in Figure 7.1.

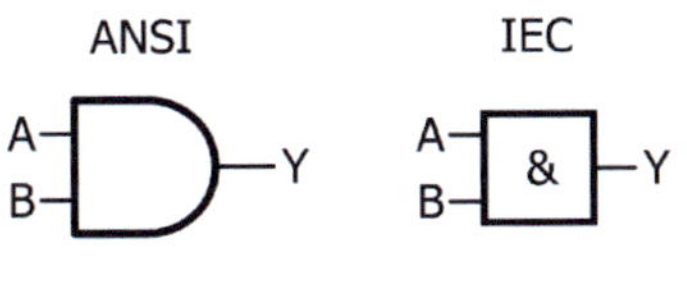

A	B	Y=AB
0	0	0
0	1	0
1	0	0
1	1	1

Figure 7.1 The AND gate's symbols, its truth table and Boolean expression.

7.2.2 OR Gate

In Boolean algebra, the OR function is the addition: FALSE + FALSE = FALSE; FALSE + TRUE = TRUE and TRUE + TRUE = TRUE (TRUE is also the maximum value, there is no such thing as 2×TRUE). In other words, the output of the OR gate is TRUE when any of its inputs is TRUE. The output is FALSE only if all its inputs are FALSE. The symbol and truth table of a two-input OR gate are shown in Figure 7.2.

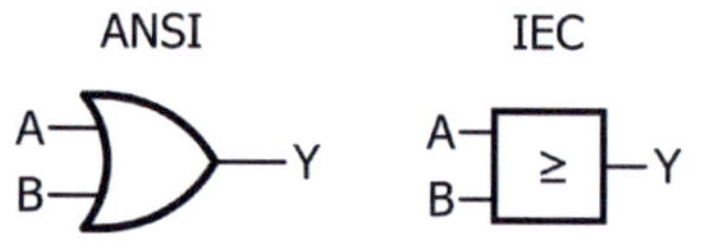

A	B	Y=A+B
0	0	0
0	1	1
1	0	1
1	1	1

Figure 7.2 The OR gate's symbols, its truth table and Boolean expression.

7.2.3 NOT Gate

The output of a NOT gate is the complement of its input. If the input is TRUE, the output becomes FALSE, and if the input is FALSE the output becomes TRUE. Therefore, the NOT gate is also called Inverter. The symbol and truth table of a NOT gate are shown in Figure 7.3. The NOT or Inverter function is often denoted with a horizontal bar above a letter or a signal name, but other symbols are use too like the '!', 'n', or '~' in front of the signal name, or single quote following the name. Inverting pins can also be drawn with a little circle at the end.

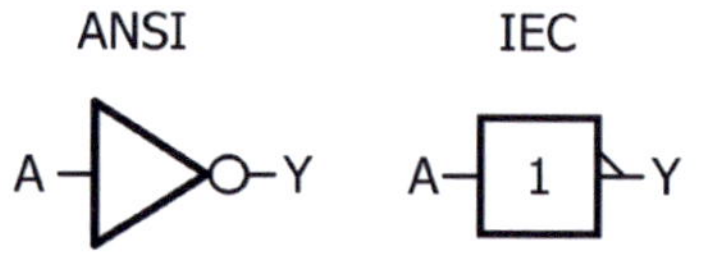

A	$Y=\overline{A}$
0	1
1	0

Figure 7.3 The NOT gate's symbols, its truth table and Boolean expression.

7.2.4 NAND Gate

The NAND gate is an AND gate followed by a NOT gate (i.e. Not AND). The output of this gate is FALSE only if all its inputs are TRUE. Figure 7.4 shows the symbol and truth table of a NAND gate.

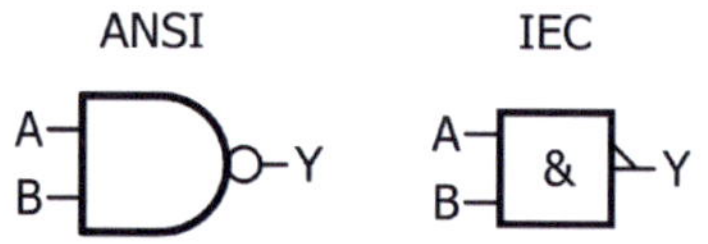

A	B	$Y=\overline{AB}$
0	0	1
0	1	1
1	0	1
1	1	0

Figure 7.4 The NAND gate's symbols, its truth table and Boolean expression.

7.2.5 NOR Gate

The NOR gate is an OR gate followed by a NOT gate (i.e. Not OR). The output of this gate is only TRUE if all its inputs are FALSE. Figure 7.5 shows the symbol and truth table of a NOR gate.

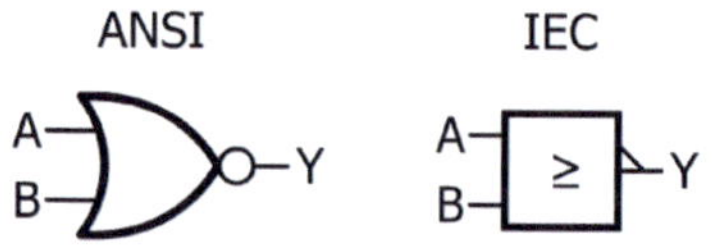

A	B	$Y=\overline{A+B}$
0	0	1
0	1	0
1	0	0
1	1	0

Figure 7.5 The NOR gate's symbols, its truth table and Boolean expression.

7.2.6 XOR gate

The output of the Exclusive OR, EXOR or XOR gate is TRUE if an odd number of its inputs are TRUE. The output is FALSE if an even number of inputs are TRUE. For a two-input XOR gate, the most common version, this means that the output is TRUE if one of its inputs is TRUE. The output is FALSE if both inputs are TRUE or when both are FALSE. Figure 7.6 shows the symbol and truth table of an XOR gate.

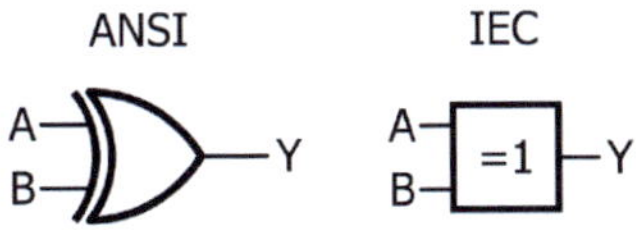

A	B	$Y=A \oplus B$
0	0	0
0	1	1
1	0	1
1	1	0

Figure 7.6 The XOR gate's symbols, its truth table and Boolean expression.

7.2.7 XNOR Gate

The Exclusive Not OR, EXNOR or XNOR is the inverted version of the XOR gate. Figure 7.7 shows the symbol and truth table of an XNOR gate.

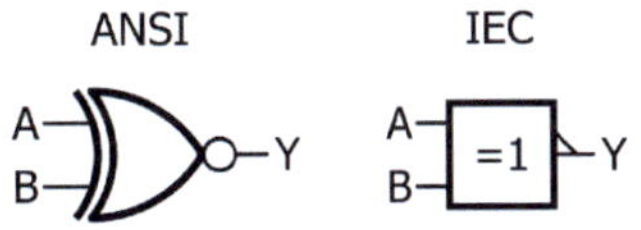

A	B	$Y=\overline{A \oplus B}$
0	0	1
0	1	0
1	0	0
1	1	1

Figure 7.7 The XNOR gate's symbols, its truth table and Boolean expression.

7.3 Logic Gate IC Families

Logic gates are available in families, depending on the technology used to manufacture them. The specifications of some commonly available families are described below.

74xx Series

The 74xx series was first manufactured in the late 1960s by Texas Instruments. There is also the 54xx series, which is the military version of the 74xx family. The ICs in this family are based on TTL (Transistor-Transistor Logic), which is why it is also known as TTL logic. The operating voltage for this family +5 V. The output current of the family members is about 16 mA. The family covers all types of logic elements, such as AND, OR, NAND, NOR, EXOR, flip-flops memories, counters, data selectors, multiplexers, etc. Some family members are:

7400: Quad 2-input NAND gate
7402: Quad 2-input NOR gate
7404: Hex inverter
7408: Quad 2-input AND gate
7410: Tripple 3-input NAND gate
7413: Dual 4-input NAND Schmitt-trigger.
7430: 8-input NAND gate
7486: Quad 2-input XOR gate

7492: Divide-by-12 counter
74161: 4-bit binary counter
74164: 8-bit parallel-out shift register
Etc., etc.

Figure 7.8 shows some example logic gate ICs. Although the 74xx series are still available, most new designs use devices based on faster or lower power and more efficient technologies (there are a large number of variants optimised for certain applications).

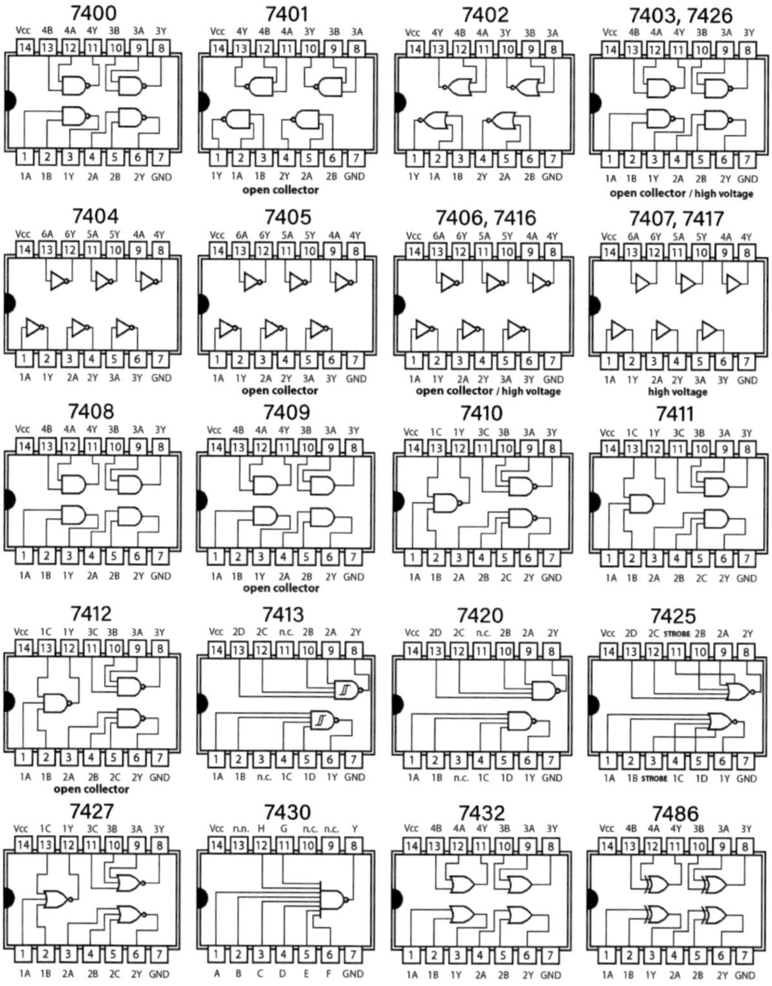

Figure 7.8 Some logic gates

The numbering system used by the 7400 family is also used by the other logic families as the standard, making it easy to identify them.

74LSxx Series
This family is also based on TTL technology but consumes less power. Because of that, it has virtually replaced the 74xx family. The operating voltage is +5 V, but the output current capability of the 74LS family is about 8 mA, which is lower than that of the 74xx family. The same numbering system as the 7400 family is used. For example, 74LS00 is a quad two input NAND gate. LS logic is common in older electronic systems.

74HCxx Series
This family is designed using high-speed CMOS technology with an operating voltage of up to +6 V. The devices in this family are pin compatible with the standard 74xx series, but the signal levels are CMOS compatible (see below), not TTL compatible. The main advantage of the 74HCxx family is the lower power consumption. Because of its CMOS signal levels, HC logic can be mixed with CMOS logic but not with TTL logic. Although an output of an HC device can drive the input of a TTL device, the opposite is not true (read: not guaranteed).

74HCTxx Series
The HCT family is like the HC family but with TTL-compatible inputs and outputs. The operation voltage is up to +5.5 V with very low-power consumption. This family is very popular as it can be mixed with both TTL and CMOS logic. The family uses the same numbering system as the standard 74xx series. For example, the 74HCT00 is a quad two-input NAND gate.

4000 Series
This is a family of logic devices based on CMOS technology. The part numbering does not follow the 74xx family standard. For example, the 4000 is a dual three-input NOR gate plus inverter. The operating voltage is up to 14 V. The 4000 family is incompatible with the TTL family and has been replaced by the HC and HCT families, the 4000 family is obsolete. Some members of the 4000 family made it into the HC and HCT because they are so practical. Think of the analogue switches (74HC(T)4016 and 74HC(T)4066) and multiplexers (74HC(T)4052 & 74HC(T)4053), and the 74HC(T)4017 decade or Johnson counter.

7.3.1 Mixing Different Logic Families

It can happen that it is necessary to mix devices from different logic families in a circuit. When doing this, two parameters are particularly important: the fan-out of a device, and the input-output voltage levels.

Fan-out refers to the number of inputs that can be driven from a single output. This feature depends on the current sink (current flowing into the output) and current source (current flowing out of the output) capabilities of the device. In general, devices from the same family can be mixed as far as the voltage levels are concerned. Examples are given below:

74xx Fan-Out
74xx devices can be freely mixed in a logic circuit since the voltage levels are compatible. The important features to consider here are:

- Low-level output current (I_{OL}) = 16 mA
- Low-level input current (I_{IL}) = 1.6 mA
- High-level output current (I_{OH}) = 0.4 mA
- High-level input current (I_{IH}) = 40 µA

We can therefore drive up to ten 74xx series inputs from one 74xx output in both the LOW state (16 mA / 1.6 mA) and the HIGH state (0.4 mA / 40 µA). It is therefore safe to drive up to ten devices from the same 74xx family. The fan-out is 10.

74LSxx Fan-Out

For the LS family, the input and output currents are as follows:

- Low-level output current (I_{OL}) = 8 mA
- Low-level input current (I_{IL}) = 0.4 mA
- High-level output current (I_{OH}) = 0.4 mA
- High-level input current (I_{IH}) = 20 µA

Therefore, the fan-out for the LS family is 20.

Mixing 74xx and 74LSxx

From the output and input specifications of both families, we see that a 74xx output can drive up to forty 74LSxx inputs. However, a 74LSxx output can drive only five 74xx inputs.

74HCTxx Fan-Out

74HCTxx family is CMOS-based, but its voltage levels are compatible with the TTL family. The 74HCTxx device input-output current specifications are:

- Low-level output current (I_{OL}) = 4 mA
- Low-level input current (I_{IL}) = 1 µA
- High-level output current (I_{OH}) = 4 mA
- High-level input current (I_{IH}) = 1 µA

Therefore, the fan-out of the HCT family is 4,000. This means that, in practice, one output can probably drive as many HCT inputs as required, since the input current is virtually zero.

Mixing LS With HCT

Because the input current of the HCT family is virtually zero, an LS output can drive thousands of HCT inputs. The other way around is more limited as an HCT output can drive only up to ten LS inputs in HIGH and LOW states.

Mixing CMOS With TTL

Mixing CMOS logic (e.g. the HC family) with TTL logic is not so easy because of the differences in voltage levels. The important voltage parameters are:

- V_{OH}: Output HIGH voltage. The minimum voltage at an output when it is logic HIGH.

- V_{OL}: Output LOW voltage. The maximum voltage at an output when it is logic LOW.

- V_{IH}: Input HIGH voltage. The minimum voltage on an input that is recognized as logic HIGH.
- V_{IL}: Input LOW voltage. The maximum voltage on an input that is recognized as logic LOW.

Table 7.1 sows the different voltage levels of TTL and CMOS inputs and outputs (assuming a +5 V supply voltage). This table shows that a logic HIGH level on a TTL output can be as low as 2.4 V. Therefore, it is not guaranteed to drive an HC (CMOS) input, which requires at least 3.5 V.

	HC	HCT	TTL
V_{OH}	4.44	2.4	2.4
V_{OL}	0.5	0.4	0.4
V_{IL}	1.5	0.8	0.8
V_{IH}	3.5	2.0	2/0

Table 7.1 Voltage levels of the TTL and CMOS gates (+5 V supply).

Figure 7.9 shows the logic voltage levels of different logic families in graphical form.

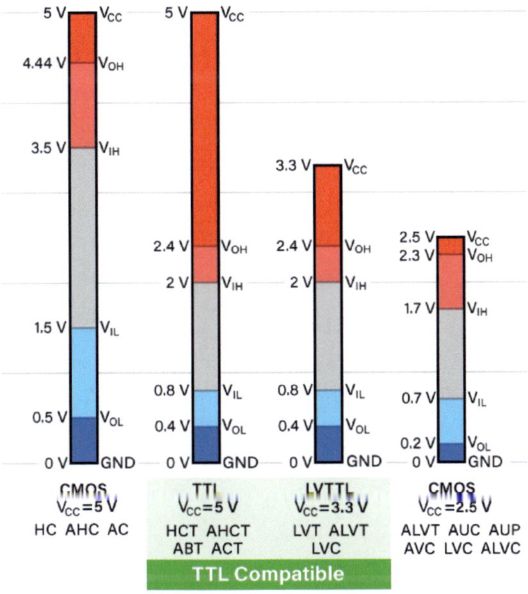

Figure 7.9 Voltage levels of different logic families.

One way to get around the TTL-driving-CMOS problem is by pulling up the output of the TTL gate with a resistor to the supply voltage, as shown in Figure 7.10. In case the two parts operate from different power supplies, a more elaborate level shifter circuit is required.

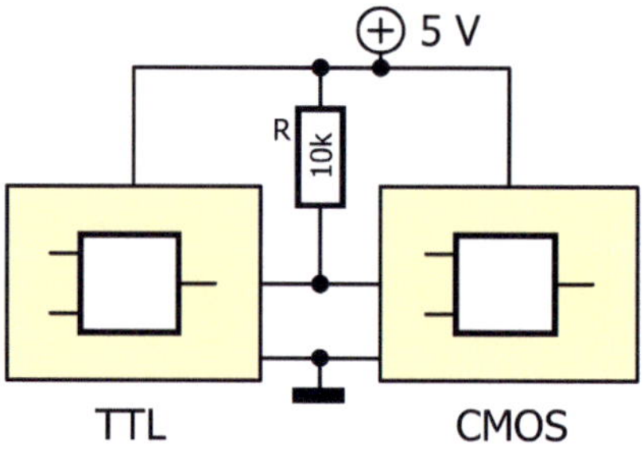

Figure 7.10 TTL-to-CMOS interface.

7.4 Project 1 – AND Gate with LED and Push-buttons

Aim: The aim of this project is to show the operation of an AND gate. Two push buttons and an LED are used. Pressing both buttons sets the output of the AND gate to logic HIGH, which turns ON the LED.

Required components:

- 180 Ω resistor
- 1 kΩ resistor
- 2× 10 kΩ resistor
- 5 mm red LED
- 5.1 V, 1 W Zener diode
- 74HCT08 AND gate
- 2× push-button
- Battery cable
- Breadboard
- Jumper wires
- 9 V battery

Circuit diagram: Figure 7.11 shows the pin configuration of the 74HCT08 quad 2-input AND gate IC. There are four AND gates inside the IC and only one of them is used in this project. The circuit diagram of the project is shown in Figure 7.12. A 5.1 V Zener diode is used to lower the 9 V battery voltage to 5 V since the operating voltage of the gate is 5 V.

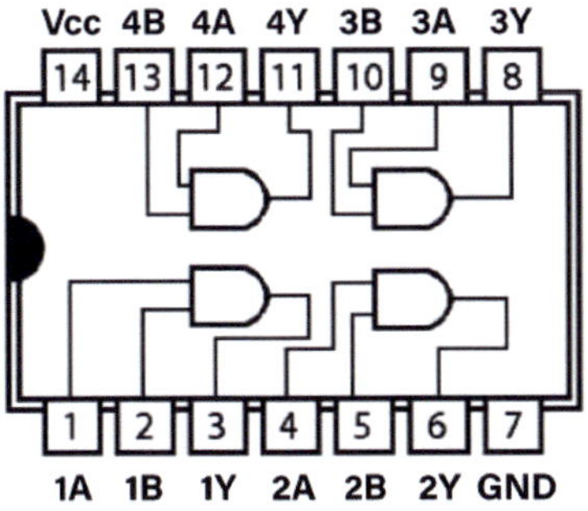

Figure 7.11 The 74HCT08 quad 2-input AND gate IC.

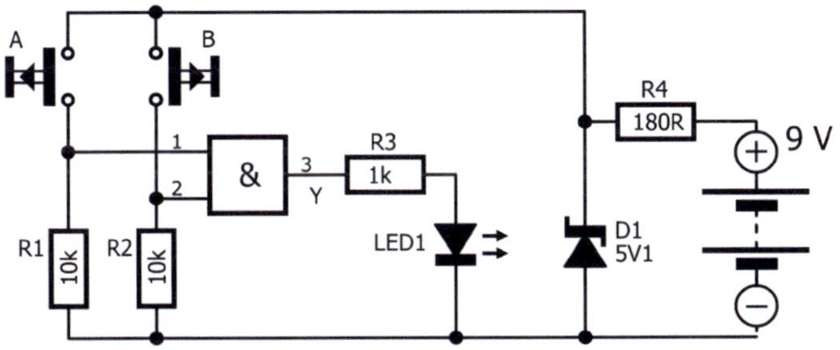

Figure 7.12 Circuit diagram of the project.

Construction: Assemble the project by placing the components on the breadboard.

Procedure:

- Connect the battery to the circuit.
- The LED should be OFF. If you press one button, the LED should stay OFF. Press both buttons at the same time to turn ON the LED.

7.5 Diode Logic

In the previous sections of this chapter, we have seen various commercially available logic gate ICs. It is interesting to know that logic gates can also be constructed using simple diodes. Examples of diode-based logic gates are presented in this section.

7.5.1 Diode OR Gate

A two-input diode OR gate consists of two diodes and a resistor, as shown in Figure 7.13. If one or both inputs are TRUE (e.g. +5 V), then current will flow through one or both diodes. As this current also flows through the resistor, it causes a voltage across the resistor, thus obtaining a TRUE on the output. When both inputs are FALSE (e.g. 0 V), the diodes will not conduct and the resistor will pull the output to FALSE.

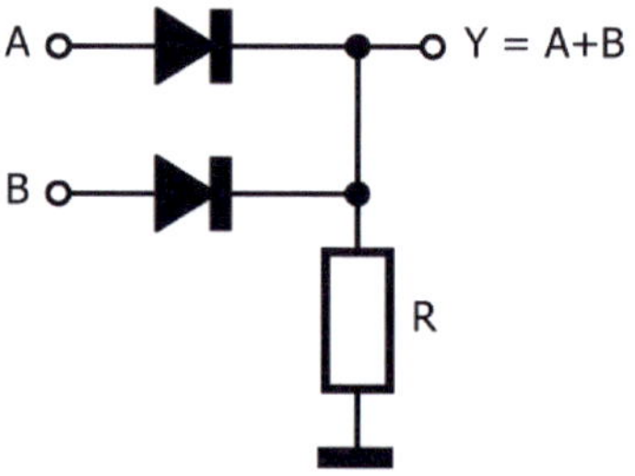

Figure 7.13 Diode-logic OR gate

7.5.2 Diode AND Gate

A two-input diode AND gate consists of two diodes and a resistor, as shown in Figure 7.14. If both inputs are at TRUE (e.g. +5 V), then the diodes are reverse biased no current can flow through them. The resistor pulls the output up to TRUE. If one of the inputs is FALSE (0 V), current flows through the corresponding diode. This current also flows through the resistor, causing a voltage drop across it and the output will be FALSE.

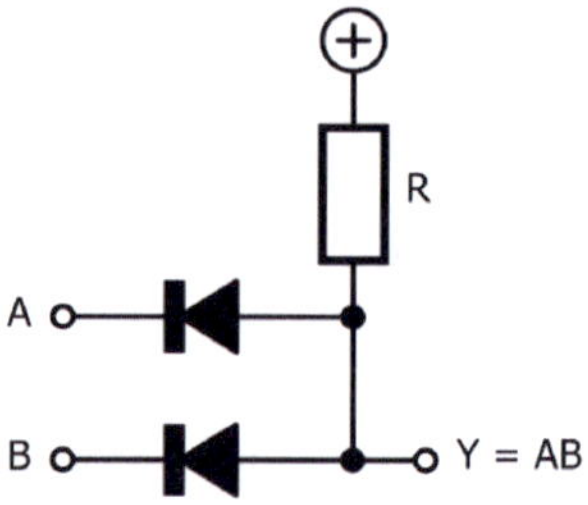

Figure 7.14 Diode-logic AND gate.

7.5.3 Mixed Diode AND & OR Gate

Figure 7.15 shows a mixed diode logic circuit with two AND gates and an OR gate.

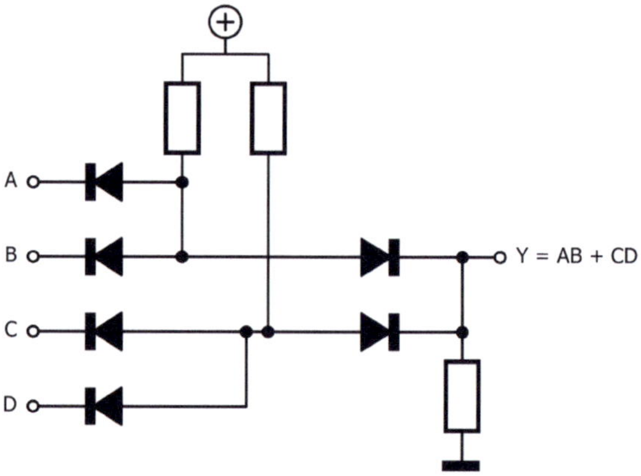

Figure 7.15 Two AND gates and an OR gate.

7.5.4 Diode Logic vs. Digital Logic

Diode-logic circuits are not suitable for complex logic circuit design. The pull-up and pull-down resistors required create relatively high output impedances. As a result of this, the output of a diode-logic gate will be affected by the input impedance on the second gate.

Diode-logic gates are also subject to voltage drop across the diodes. This voltage drop problem gets worse when several stages are connected in series, and the circuit becomes unusable.

Diode-logic gates also have high power consumption compared to digital logic devices, such as TTL or CMOS. This is a serious limitation since complex diode-logic circuits will dissipate large amounts of power.

Diode-logic gates also have slow switching speeds, which limits their usefulness in high-speed designs.

Diode logic also has some advantages. They can be useful when only one or two gates are needed in a circuit, saving an IC. They can provide level shifting thanks to the pull-up or pull-down resistor. Also, they can work on much higher voltages than digital logic. Some diodes can support hundreds of volts. Another interesting property is that a diode-logic gate may have different voltage ranges on its inputs. This property is often used in circuits with multiple power supply inputs. A diode OR gate (without the pull-up resistor) will automatically select the highest available voltage.

7.5.5 Single-Gate Logic Devices

Single-gate logic devices (like Little Logic or TinyLogic), also known as the 1G series, contain a single gate, such as NAND, NOR, AND, OR, and EXOR gates with two or more inputs. They offer very low working voltages down to 0.9 V and are often used in battery-operated devices. Such gates with two inputs come in tiny five-pin packages, making them somewhat complicated to use in hobby-type electronics work.

7.6 Flip-Flops

Digital flip-flops are circuits which have two stable states that can be used to store binary data. By applying a trigger or clock signal, one can change the state of the stored data from one state to the other. Flip-flops are widely used in digital counters, computers, microcontrollers, in digital communications systems, and in digital control systems.

Flip-flops are available as integrated circuits. Some common flip-flop types are:

- SR flip-flop
- JK flip-flop
- D flip-flop

SR Flip-Flop

'SR' stands for Set/Reset, which are the inputs of the flip-flop. The Set input sets the output to TRUE whereas the Reset input sets the output to FALSE.

An SR flip-flop is made up of two cross-coupled NAND or NOR gates. Figure 7.16 shows an SR flip-flop made from NAND gates.

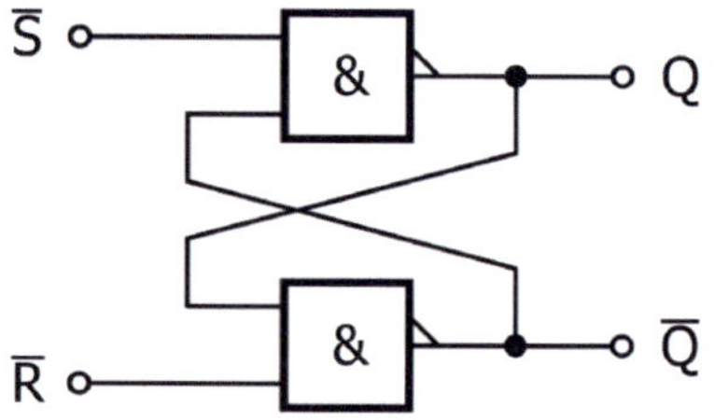

Figure 7.16 NAND-gate-based SR flip-flop

The SR flip-flops has two inputs S (Set) and R (Reset) and two outputs Q and Q' (not Q, the opposite of Q). When S becomes TRUE while R remains FALSE, the output Q will become TRUE (i.e. 1). When S returns to FALSE, the output remains TRUE. When R becomes TRUE while S is FALSE, the output Q will become FALSE (i.e. 0). When R returns to FALSE, the output remains FALSE. The truth table is shown in Table 7.2.

S	R	Q	Q'
0	0	0	1
0	1	0	1
1	0	1	0
1	1	Invalid	Invalid

Table 7.2 SR flip-flop truth table

JK Flip-Flop

JK are the initials of Jack Kilby, co-inventor of the integrated circuit. JK can also mean Just Kidding. In both cases, there is no relation with the function of the signals.

The JK flip-flop is an improvement on the SR flip-flop that adds a clock signal. It avoids the invalid state S=1, R=1. Figure 7.17 shows a JK flip-flop. J is like S and K is like R. However, when both are TRUE, the output Q will toggle for every clock signal instead of being undefined. When J and K are both FALSE, the output Q doesn't change.

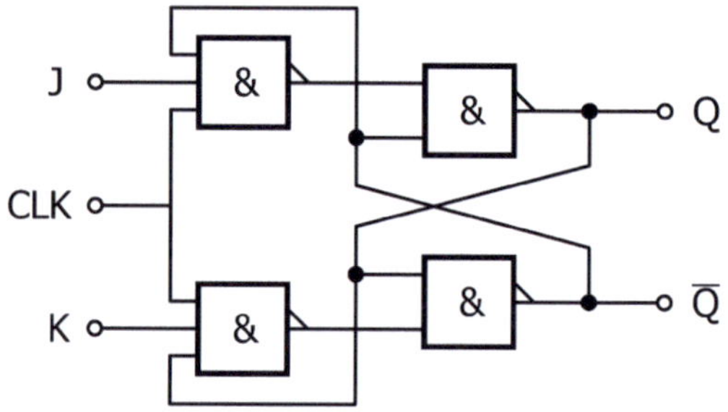

Figure 7.17 NAND-gate-based JK flip-flop.

When J = K = 1 and CLK goes from low-to-high, then the output changes state. If J and K inputs are different, the output takes the value of J. If both J and K are low, then no change occurs in its output. The truth table is shown in Table 7.3. Qn is the present state of the output, and Q(n+1) is the next state.

CLK	J	K	Q(n+1)
Low-to-high	0	0	Qn
Low-to-high	0	1	0
Low-to-high	1	0	1
Low-to-high	1	1	Qn'

Table 7.3 JK flip-flop truth table

D Flip-Flop

The Data or D flip-flop is commonly used in shift registers, counters and memories. Figure 7.18 shows the circuit diagram of the D flip-flop. It is like a JK flip-flop where D is J and where K is not J (J'). Therefore, on every rising edge (low-to-high transition) of the clock signal the output Q takes on the value of D. The truth table of the D flip-flop is shown in Table 7.4.

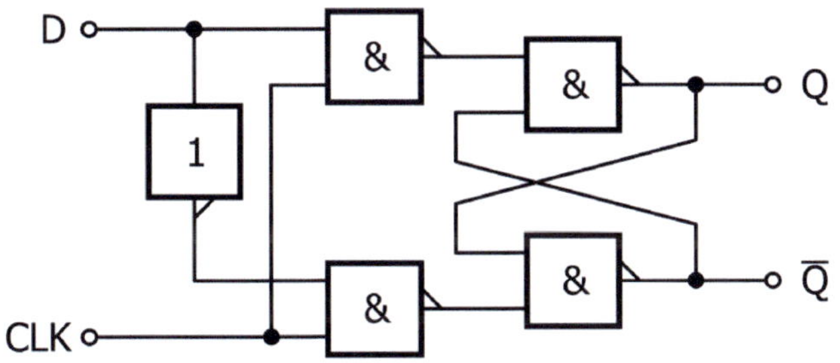

Figure 7.18 NAND-gate-based D flip-flop.

CLK	D	Q	Q'
0	0	0	1
Low-to-high	0	0	1
Low-to-high	1	1	0

Table 7.4 JK flip-flop truth table

7.7 Counters

Digital counters are devices with 4 or more outputs (bits) that count up and/or down in binary when clock pulses are applied to their CLK inputs. (There also exist variants that count in decimal a.k.a. BCD, which stands for Binary-Coded Decimal.) Digital counters are made using flip-flops, but their design is beyond the scope of this book. In this section, we will look at the two types of counters and then look at the details of a counter IC.

Asynchronous counters: In these counters the clock pulse is applied to the first flip-flop and then the output of a flip-flop is used as the clock for the next stage.

Synchronous counters: In these counters, the clock pulse is applied to all the flip-flops at the same time.

A comparison of asynchronous and synchronous counters is shown in Table 7.5.

Synchronous counter	Asynchronous counter
Clock is applied to all flip-flops at the same time	Clock is applied to the first flip-flop only
Faster	Slower
Design is difficult	Design is easier
Can operate in any order	Only up/down counting
Small propagation delay	Large propagation delay

Table 7.5 Synchronous counters vs. asynchronous counters

As an example, the 74161 IC (also available as 74HC161, 74HCT161) is a 4-bit synchronous counter chip. The device counts from 0 to 15 (0b0000 to 0b1111 in binary notation, or from 0x0 to 0xf in hexadecimal notation, see table 7.6) and then starts over. The 74161 has a user-controllable asynchronous clear function. Thus, the outputs of the counter are cleared to 0 as soon as the clear input (CLR) is activated, independent of the CLK input.

The counter can be preloaded to start from any value. Multiple chips can be cascaded (connected in series) to make a wider counter.

Applications of this counter include frequency counters and dividers, digital clocks, memory address generators, etc. Figure 7.19 shows the pin configuration of the 74161.

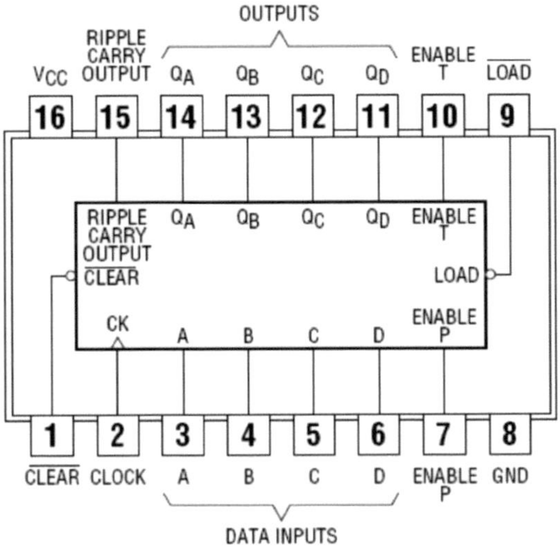

Figure 7.19 The 74161 4-bit synchronous counter.

The pin definitions of the 74161 are:

QA, QB, QC, QD:	4-bit counter output.
A, B, C, D:	4-bit input value that can be loaded into the counter.
LOAD:	Setting this pin LOW loads the data at A, B, C, D into the counter.
CLK:	Clock pulse input to advance the counter.
CLR:	Setting this pin LOW clears the outputs to 0.
GND, VCC:	Power supply pins.
RCO:	Ripple carry output indicates that the counter went from 15 to 0 (overflow).
ENP, ENT:	Count-enable inputs. Must be HIGH to allow counting.

Clock pulses must be applied to the clock input of the counter to advance it. Clock pulses can be generated by, for example, a multivibrator, a 555 chip, or some other method. The counting speed depends on the frequency of the clock pulses. When clock pulses are applied, the outputs count as shown in Table 7.6.

QD	QC	QB	QA	decimal	hexadecimal
0	0	0	0	0	0
0	0	0	1	1	1
0	0	1	0	2	2
0	0	1	1	3	3
0	1	0	0	4	4
0	1	0	1	5	5
0	1	1	0	6	6
0	1	1	1	7	7
1	0	0	0	8	8
1	0	0	1	9	9
1	0	1	0	10	a
1	0	1	1	11	b
1	1	0	0	12	c
1	1	0	1	13	d
1	1	1	0	14	e
1	1	1	1	15	f

Table 7.6 Output of a 4-bit binary counter.

Looking at the Qx columns individually, you'll notice that QA is equal to the clock frequency divided by two, QB is the clock frequency divided by four, QC is the clock frequency divided by eight, and QD is the clock frequency divided by sixteen. Therefore, a counter can be used as a frequency divider. Figure 7.20 shows Table 7.6 as a timing diagram.

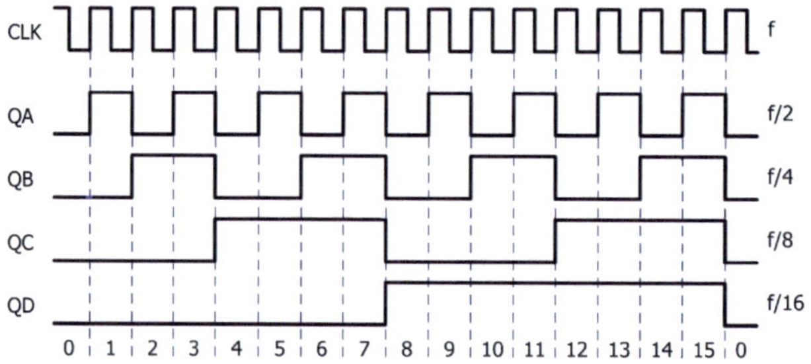

Figure 7.20 Timing diagram of a 4-bit counter.

7.8 The 74595 Shift Register

The 74595 (also available as 74HC595, 74HCT595) is a popular serial-in, serial- or parallel-out shift register. The operation of a shift register is such that serial data is clocked into the device at every clock pulse. This data is output in parallel form at the output of the shift register. In summary, this is what happens at each clock pulse:

- The bits in the shift register are shifted by one position, i.e. bit 0's value is shifted into bit 1, bit 1's value is shifted into bit 2 and so on.

In the case of the '595 a latch or store pulse must be provided to make the data available at the parallel outputs of the shift register. This pulse can be provided after each clock pulse or after several clock pulses.

Figure 7.21 shows the pin configuration of the 74595 shift register.

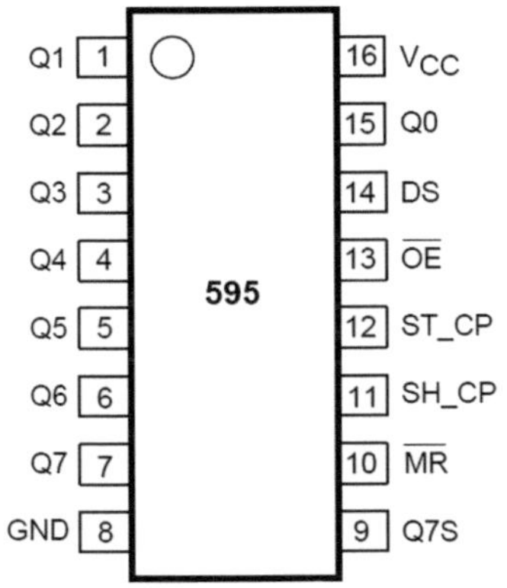

Figure 7.21 The popular 74(HC(T))595 shift register.

The pin definitions of the '595 chip are:

DS:	Serial input.
SH_CP:	Clock input.
MR:	Master reset to clear all shift register data to 0.
OE:	Output enable. LOW to enable output data.
ST_CP:	Pulse to transfer the data to the output.
Q0 to Q7:	Shift register parallel output.
Q7S:	Serial output (for cascading multiple devices)
GND, VCC:	Power supply pins

Some application areas of shift registers are:

- Serial to parallel data conversion.
- Capture and hold data for extended periods of time.
- In industrial control.
- In computer peripherals.
- To minimize microcontroller pin usage.

An application of an 8-bit shift register could be to control eight LEDs from a single data output as shown in Figure 7.22. The LEDs are connected to the outputs of the shift register. Eight bits of data could be shifted into the shift register corresponding to the eight LEDs such that 1s are sent for the LEDs to be turned ON and 0s for the LEDs to be turned OFF.

For example, shifting in the bit pattern '01110001' will turn ON the LEDs at bit positions 0, 4, 5, and 6 (bit positions are counted starting at the right with 0, i.e. bit 0 is on the right).

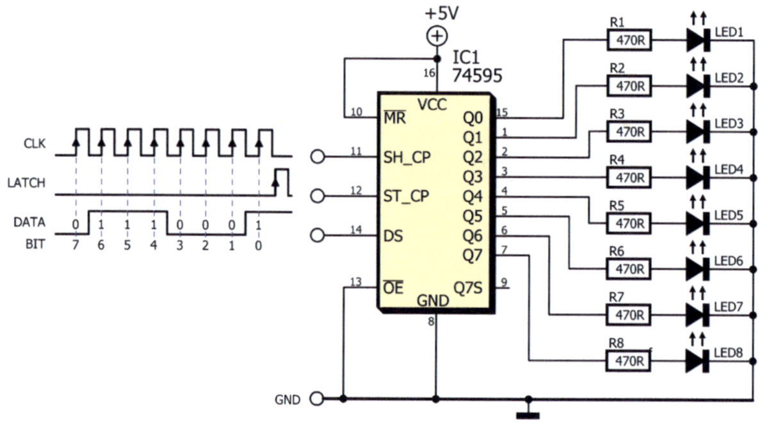

Figure 7.22 Using a '595 shift register to control eight LEDs from a single-bit data input.

7.9 What is an FPGA?

In this book we will not use FPGAs, but it is good to know that they exist and what they are. A Field-Programmable Gate Array or FPGA is a reconfigurable logic device containing many thousands of gates and flip-flops. It can be programmed or reprogrammed if required after manufacturing. At the core of an FPGA is an array of programmable logic blocks that are interconnected such that they can be programmed to perform various digital functions. Because of their programmable nature, FPGAs are ideal for the design of complex digital systems. Some application areas of FPGAs are:

- Automotive industry
- Consumer electronics
- Industrial controls
- Aerospace and defence
- Data centres
- High-speed computing
- Data storage
- Medical
- Video processing
- Image processing
- Wireless communications, etc.

Designing complex digital circuits using logic gates is a highly complex task and has many disadvantages like consuming a lot of space and energy. Perhaps the biggest disadvantage is that such a design is fixed and cannot easily be changed. FPGAs, on the other hand, are reconfigurable and flexible. They are especially useful for building fast and highly complex digital circuits.

Advantages of FPGAs

- The functionality can easily be changed since they are programmable.
- Much smaller board space is required. Designing a complex digital circuit using gates may require several hundred or more gates, and this may occupy large area. An FPGA, on the other hand, consists of a single chip.
- Designing a complex digital circuit using FPGAs is much easier than using gates thanks to special tools and programming languages like VHDL and Verilog.
- Development costs are relatively low.
- The power consumption can be low.

Chapter 8 • Advanced Subjects

8.1 Overview

In this chapter, we will be studying advanced topics such as Alternating Current (AC) circuits, filters, oscillators, amplifiers, commonly used sensors and so on.

8.2 Alternating Current (AC) Analysis

AC voltage is like a sine wave that periodically reverses its direction and its magnitude. The mains electricity you use at home is in AC form. Figure 8.1 shows an AC voltage. In this figure, Vo is the peak voltage and T is the period. For example, in the UK, the peak voltage is ±325 V and the period is 20 ms. It is common practise to specify the frequency of an AC voltage rather than its period. For example, in the UK, the frequency of the mains AC voltage is:

$$f=\frac{1}{T}$$

Which gives, f = 50 Hz

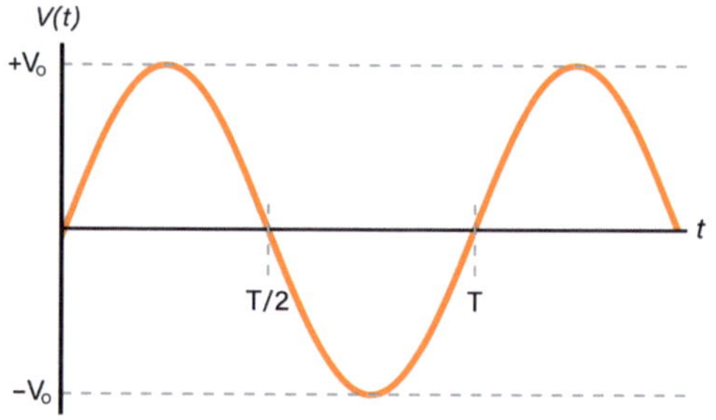

Figure 8.1 An AC voltage waveform.

The AC voltage is represented mathematically as:

$$V(t)=V_O\sin(\omega t)$$

Where V_O is the peak voltage and ω is the radian frequency:

$$\omega=2\pi f$$

For example, the mains voltage in the UK can be expressed as:

$$V(t)=325\cdot\sin(2\pi\cdot 50t)$$ or

$$V(t)=325\cdot\sin(314t)$$

Inspecting the above equation, we can see that the voltage varies between +325 V and −325 V with a frequency of 50 Hz.

Wait a minute. Isn't the mains voltage 230 V? It is, we'll come to that right now.

8.2.1 Root Mean Square (RMS)

In AC analysis, we usually refer to the RMS value of a voltage. The RMS value of an AC voltage is a measure of the equivalent DC voltage that would produce the same heating effect in a resistive load as the AC voltage. The RMS value is calculated using the following formula:

$$V_{RMS}=\frac{V_O}{\sqrt{2}}=\frac{V_O}{1.41}=0.707\cdot V_O$$

For example, in the UK, the RMS voltage of the mains is equal to 325 × 0.707 = 230 V. This means that if we connect an electric heater to a 230 V DC supply, then the same power will be dissipated by the heater as if it is connected to the mains voltage. When we say that the mains voltage is 230 V, we actually refer to its RMS value.

8.2.2 Resistors and AC

A resistor behaves the same way for an AC source as for a DC source. Therefore, the current through a resistor R in an AC circuit is given by the following equation.

$$i(t)=\frac{V_O\sin(\omega t)}{R}$$

The current and the voltage are in phase with each other.

8.2.3 Inductors and AC

In an AC circuit, an inductor presents resistance called the inductive reactance. The inductive reactance is denoted by X_L, where $X_L = \omega L$. The unit of inductive reactance is in ohms. Note how X_L is proportional to ω. Therefore, the resistance of an inductor is proportional to frequency and is called impedance, denoted by Z.

The current through the inductor is given by the following equation:

$$i(t)=\frac{V_O}{X_L}$$

or,

$$i(t)=\frac{V_O}{\omega L}$$

Where L is the inductance in henries.

In such a circuit, the voltage and current are not in phase. The voltage *leads* the current by 90º.

8.2.4 Capacitors and AC

A capacitor does not block AC signals but acts like a resistor. This resistance is called the capacitive reactance and is denoted by XC, where $X_C = 1/\omega_C$. Like the inductive reactance, the unit of capacitive reactance is also in ohms. Note how X_C is proportional to ω. Therefore, the resistance of a capacitor is proportional to frequency and is called impedance, denoted by Z.

The current through the capacitor is given by the following equation:

$$i(t)=\frac{V_O}{X_C}$$

or, $$i(t)=V_O\omega C$$

Where C is the capacitance in farads.

In such a circuit, the voltage and current are not in phase and the voltage lags the current by 90º.

8.2.5 RLC Circuit and Resonance

RLC circuits are very common in many electronic circuit applications. The resistance R usually represents the DC resistance of the inductor. In general, its value is very low, and it can be ignored. In this section, we will investigate the response of an RLC circuit and discuss the important topic of resonance in such a circuit.

Figure 8.2 shows a simple series RLC circuit with an AC voltage $V(t) = V_O \sin(\omega t)$ applied to the circuit.

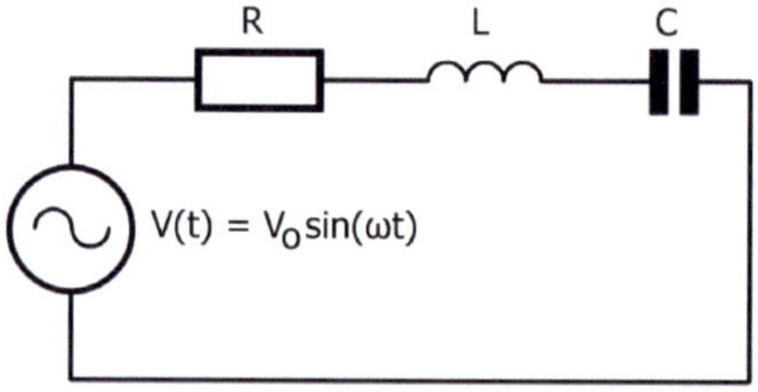

Figure 8.2 RLC circuit with AC input voltage.

Impedance, denoted by Z, is the AC resistance of a circuit. It can be shown that the total impedance of the above circuit is given by:

$$Z=\sqrt{R^2+(X_L-X_C)^2}$$

or, $$Z=\sqrt{R^2+\left(\omega L-\frac{1}{\omega C}\right)^2}$$

Resonance is defined to occur when ωL = 1/ωC. When in resonance, the impedance of the circuit is at a minimum and equal to R. The resonance radian frequency is:

$$\omega=\frac{1}{\sqrt{LC}}$$

The resonance frequency in hertz is,

$$f=\frac{1}{2\pi\sqrt{LC}}$$

At resonance, the current in the circuit reaches a maximum (Figure 8.3) given by

$$i(t)=\frac{V_O}{R}$$

The result of this is that the voltage across the inductor and the capacitor may become many times larger than the supply voltage, depending on the value of R.

Resonance is used in tuned circuits to select a certain frequency. For example, in radio antenna circuits, the frequency that puts the circuit into resonance has the highest amplitude and is the selected frequency. By adjusting the value of the capacitor (adjusting an inductor is more complicated), the selected (resonance) frequency changes. This allows selecting different stations by the radio.

The amount of this current (or voltage) amplification is shown by the Q factor of the resonant circuit and is given by:

$$Q=\frac{\omega L}{R}=\frac{1}{\omega RC}$$

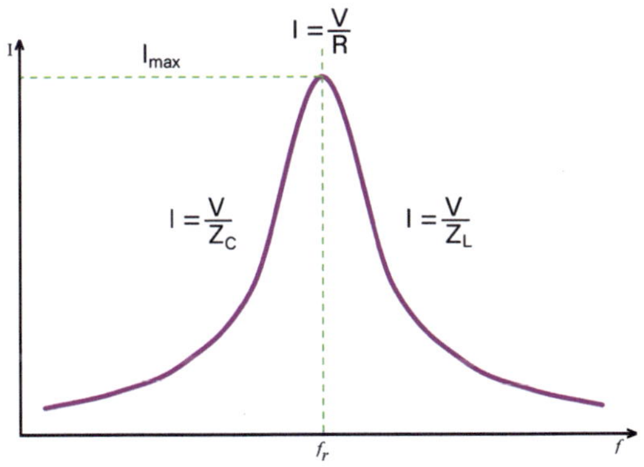

Figure 8.3 Circuit current at resonance.

The Q factor determines the sharpness or the selectivity of a resonance circuit. The Q factor is increased by lowering the DC resistance R of the inductor, which depends largely on the geometry of the coil winding.

The selectivity of a resonant circuit determines the bandwidth of the circuit. The sharper the circuit selectivity, the smaller is the bandwidth. Bandwidth is defined by two frequencies above and below the resonant frequency (a.k.a. center frequency) where the response is ½√2 = 70.7% of the circuit's peak value. These two frequencies are also known as the −3 dB points. The bandwidth BW is related to the Q factor by the expression:

$$Q=\frac{f_r}{BW}$$

Where, fr is the resonant frequency in Hz, and BW is the bandwidth in Hz (Figure 8.4). As explained above, for a high selectivity, we require a high Q factor, and therefore a small bandwidth.

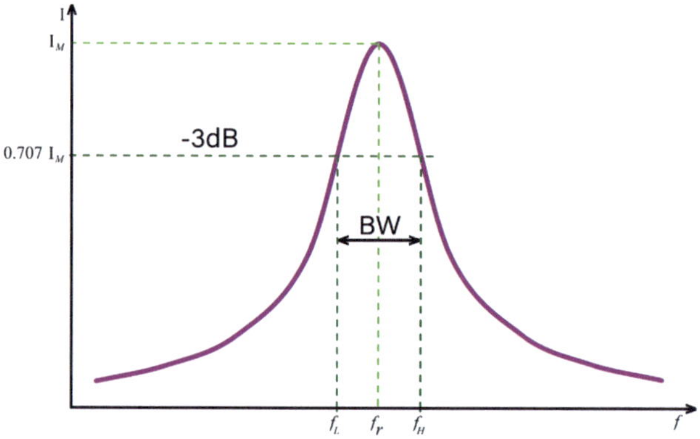

Figure 8.4 The bandwidth.

Example
It is required to calculate the inductor value in a resonant circuit. The required bandwidth is 100 kHz, and the resonant frequency is 1 MHz. Assume that the DC resistance of the inductor is 2 Ω.

Solution
From the equations,

$$Required\ Q=\frac{1\cdot 10^6}{1\cdot 10^3}=10$$

Also,

$$Q=\frac{\omega L}{R}$$

Rearranging the equation,

$$L=\frac{QR}{\omega}$$

Where ω = 2πf. So, ω = 2π × 1 × 10^6 = 6.28 × 10^6 rad/s.

Therefore, L = 10 × 2 / (6.28 × 10^6) giving L = 3.18 × 10^{-6} H or L = 3.18 μH.

Simulation

Here, we will simulate the voltage across an inductor and capacitor in a series RLC circuit when the circuit is in resonance. We will assume that L = 10 μH, C = 100 pF and the supply voltage is 10 V. Also, assume that the DC resistance of the inductor is 50 Ω.

From the resonance formula, the resonant frequency of the circuit is:

$$f=\frac{1}{2\pi\sqrt{LC}}$$

Which gives, f = 5,032,921 Hz and ω = 31,622,777 rad/s. The Q factor is ωL/R, giving Q = 6.3. We therefore expect the voltage across the inductor to be 6.3 times the supply voltage (i.e. 63 V) when the circuit is in resonance.

The steps are as follows:

- Start the TINA-TI simulator program.
- Click on resistor, capacitor, and inductor and connect them in series as in Figure 8.5. Set the values of the components by right-clicking on them.
- Click on Voltage Generator and insert it to the circuit. Set its frequency to 5033 kHz and its amplitude to 10 V.
- Insert an earth symbol at the bottom part of the circuit. Every circuit to be simulated must have a path to earth.
- Click **Analysis → AC analysis → Table of AC results**
- You should see all the AC voltages and currents displayed in a table. Notice that the voltage across the inductor (V_L[3,2]) = 63.25 V as expected. Also, the voltage across the capacitor (V_C[1,3]) = 63.25 V with -180° phase shift.

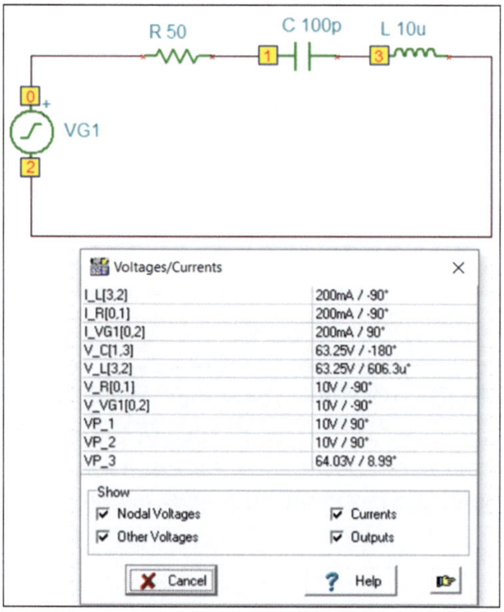

Figure 8.5 The RLC circuit built in the simulator.

The bandwidth of the circuit in this example is: BW = f_r / Q = 5,032,921/6.3 = 798,876 Hz.

8.2.6 Power in AC circuits

The Instantaneous Power in an AC circuit is calculated by multiplying the instantaneous voltage with the instantaneous current:

$$P = V_{inst} \times I_{inst}$$

One of the important power equations is the Real Power, also called the true power. Real power is the average power dissipated by a resistive load as if the power came from a DC source. The unit of real power is watts. It is calculated by multiplying the RMS values of the voltage and current with the cosine of the phase angle between the voltage and current:

$$P_{real} = V_{RMS} \times I_{RMS} \times \cos(\theta)$$

cos(θ) is known as the Power Factor of the circuit, and it indicates the efficiency of an AC circuit.

There is also Apparent Power, which is calculated by multiplying the RMS values of the voltage and current:

$$P_{apparent} = V_{RMS} \times I_{RMS}$$

The unit of apparent power is volt-ampere, i.e. VA.

Example

Consider an AC circuit with the following parameters. Calculate the various power values of this circuit:

V_{RMS} = 200 V
I_{RMS} = 10 A
Phase angle = 60°

Solution

From the above equations:

Apparent power = 200 × 10 = 2000 VA
Real power = 200 × 10 × cos(60) = 1000 W
Power factor = cos(60) = 0.5

The power factor indicates that the circuit is operating at 50% efficiency. Only half of the power is converted into useful work.

8.3 Amplifiers

Amplifiers are electronic circuits which increase the level of their input signals. Amplifiers are classified according to the size and the frequency of the signals they handle. Small signal amplifiers are used to amplify small signals like the output of a microphone, a sensor or a receiver so they can be used for further processing. Power amplifiers (or large signal amplifiers) are used to amplify signals so they can drive a load like a loudspeaker, a transducer or a radio transmitter.

8.3.1 Amplifier Gain

The amplifier gain is the ratio of its output to its input. Gain has no units. There are three types of amplifier gains: voltage gain, current gain, and power gain. In reference to Figure 8.6, we can define the three types of gains as follows.

Figure 8.6 The amplifier as a black box.

Voltage Gain

This is the ratio of the amplifier's output voltage to its input voltage. Expressed mathematically:

$$\text{Voltage gain } A_V = \frac{output\ voltage}{input\ voltage} = \frac{V_o}{V_i}$$

The voltage gain is usually expressed in decibels (dB) using the following relationship:

$$\text{Voltage gain in dB} = 20 \cdot log_{10}\,(\frac{V_o}{V_i})$$

or, $$\text{Voltage gain in dB} = 20 \cdot log_{10}\,(A_V)$$

Where log is the logarithm with base 10. An example is given below.

Example
The output and input voltages of an amplifier are 2 V (2,000 mV) and 10 mV respectively. Calculate the voltage gain of the amplifier both as a ratio and in dB.

Solution

As a ratio:	A_V = 2000 / 10 = 200
In dB:	20 × $\log_{10}(200)$ = 46 dB

Current Gain
This is the ratio of the amplifier's output current to its input current. Expressed mathematically:

$$\text{Current gain } A_i = \frac{output\ current}{input\ current} = \frac{I_o}{I_i}$$

The current gain is usually expressed in decibels (dB) using the following relationship:

$$\text{Current gain in dB} = 20 \cdot log_{10}\,(\frac{I_o}{I_i})$$

or, $$\text{Current gain in dB} = 20 \cdot log_{10}\,(A_i)$$

Example
The output and input currents of an amplifier are 500 mA and 10 mA respectively. Calculate the current gain of the amplifier both as a ratio and in dB.

Solution

As a ratio:	A_i = 500 / 10 = 50
In dB:	20 × $\log_{10}(50)$ = 34 dB

Power Gain
This is the output power of an amplifier divided by its input power. Expressed mathematically:

$$Power\ gain\ A_P = A_V\, A_i$$

Expressed in decibels (note the factor 10 instead of 20):

$$\text{Power gain in dB} = 10 \cdot log_{10}\,(A_P)$$

Efficiency

The efficiency of an amplifier is the ratio of the power it outputs to the power it consumes. Expressed mathematically:

$$\text{Efficiency} = \frac{output\ power}{input\ power} \cdot 100\%$$

8.3.2 The LM386 Amplifier

In the past, electronics engineers and hobbyists used to design audio amplifiers using discrete components such as transistors, resistors, and capacitors. Nowadays, there are many audio amplifier ICs that can be used to design audio amplifiers and not many engineers design transistor-based audio amplifiers anymore. The advantage of using an IC is that it requires only a few external components, resulting in a simpler circuit. In addition, the IC contains a proven design and technology, and it should work as long as the external components are dimensioned properly and connected correctly.

The LM386 is one of the most popular low-cost and low-power audio amplifier ICs designed to replace transistor-based audio amplifier. In this section, we will learn the basic features of this IC.

The LM386 is a mono low-voltage amplifier that can be used in various applications. It can drive loads from 4 Ω up to 32 Ω. The gain is internally set to 20, but it can be modified from 20 to 200 by placing a resistor and capacitor between its pins 1 and 8. The device comes in different 8-pin packages such as PDIP, SOIC and VSSOP to fit different applications.

Features of the LM386 amplifier:

- Battery operation
- Few external components
- Wide supply voltage (4 V – 12 V for the LM386N, or 5 V – 18 V for the LM386N-4)
- Low current (4 mA)
- Voltage gain from 20 to 200
- Low distortion (0.2%)
- 250 mW output power in 8 Ω
- 8-pin package

Figure 8.7 shows the pin configuration of the LM386. The pin definitions are shown in Table 8.1.

Pin		Type	Description
Nr	Name		
1	Gain	–	Gain setting pin
2	–Input	I	Inverting input
3	+Input	I	Noninverting input
4	GND	P	Ground reference
5	VOUT	O	Output
6	VS	P	Power supply voltage
7	Bypass	O	Bypass decoupling path
8	Gain	–	Gain setting pin

Table 8.1 LM386 pin definitions. I = input, O = output, P = power.

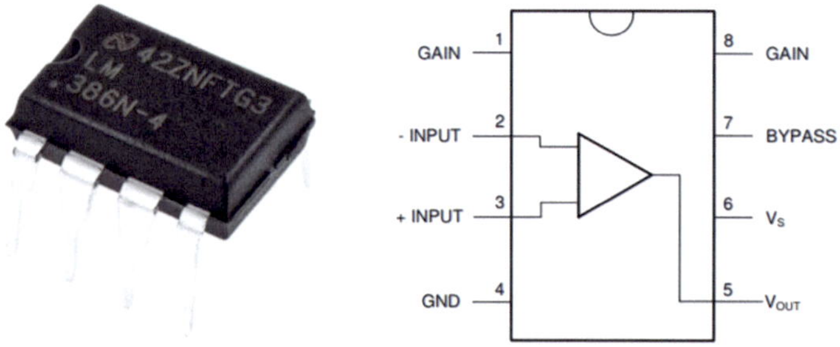

Figure 8.7 LM386 pin configuration

Figure 8.8 shows the circuit diagram of an audio amplifier with volume control based on an LM386 with its voltage gain set to 200 (46 dB). The input signal is applied to pin 3 through a potentiometer (P1), which acts like a volume control. Pin 6 should be connected to the positive terminal of a power supply, e.g. a 9 V battery, while pin 4 must be connected to the negative terminal (ground). The capacitor between pin 1 and 8 (C1) sets the voltage gain to 200. Removing this capacitor will set the voltage gain to 20. A 10 μF capacitor is recommended between pin 7 and ground to eliminate noise. The output signal is taken from pin 5 through a coupling capacitor (C2) and drives an 8 Ω loudspeaker. The power supply should be well regulated, and the voltage must be within the specified values. It is recommended to place a capacitor to GND close to the LM386 power supply pin.

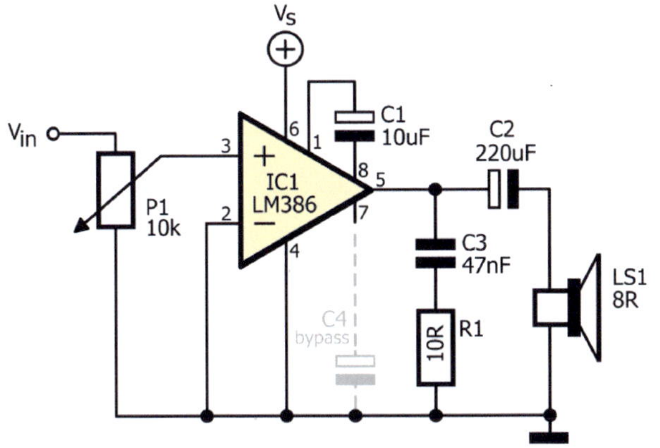

Figure 8.8 Schematic of a simple LM386-based audio amplifier. C4 is optional.

8.3.3 The TDA2030

Another common amplifier IC is the TDA2030, is an 18 W audio power amplifier having the following features:

- 18 W output into a 4 Ω loudspeaker
- ±16 V supply voltage
- High output current (60 mA)
- Very low harmonic and crossover distortion
- Short-circuit protection
- Thermal shutdown

Figure 8.9 shows the TDA2030 together with the circuit diagram of a powerful audio amplifier based on this IC. Further details on the TDA2030 IC can be obtained from the manufacturer's data sheets.

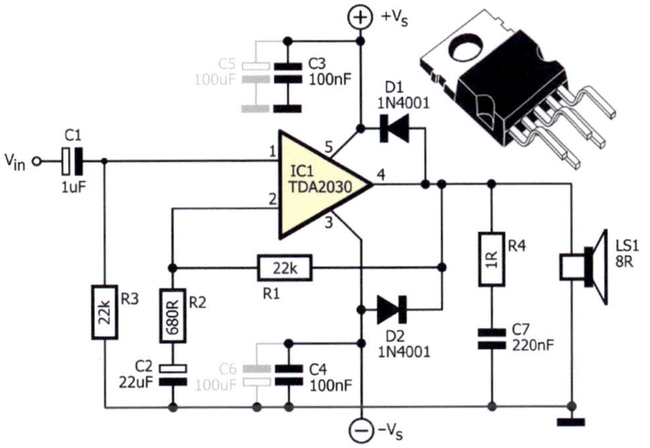

Figure 8.9 The TDA2030 IC and audio amplifier. C5 and C6 are optional.

8.4 Oscillators

In earlier chapters, we have seen how to design transistor-based sine wave oscillators and square wave multivibrators. In this section, we will design oscillator circuits using operational amplifiers. Simulation of the circuits will be given for each design.

8.4.1 Phase-Shift Sine Wave Oscillator

Figure 8.10 shows a phase-shift oscillator designed using an operational amplifier. The output waveform is also shown in the same figure. Here, the resistor-capacitor network provides the required 180° phase shift for the feedback network. Assuming C1 = C2 = C3 and R2 = R3 = R4, the frequency of oscillation is given by:

$$f = \frac{1}{2\pi RC\sqrt{6}}$$

The condition for oscillation is that the voltage gain of the operational amplifier must be greater than 29. With the component values chosen in Figure 8.10 (R = 6.8 kΩ, C = 10 nF), the oscillation frequency is f = 1 kHz. The voltage gain of the circuit is R1 / R2. The condition for oscillation is satisfied if R1 / R2 > 29. Since R1 = 6.8 kΩ, we can choose R2 = 250 kΩ (220 kΩ would be a suitable E12 value).

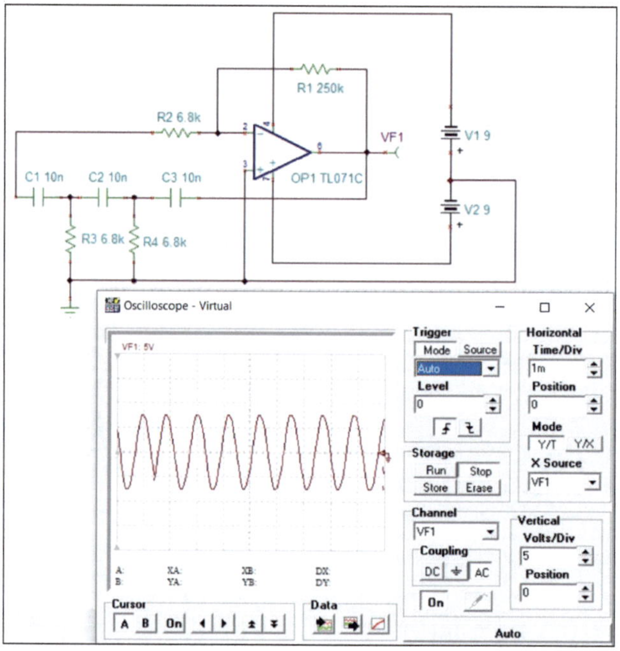

Figure 8.10 Phase shift oscillator.

Procedure:

- Start the TINA-TI simulator.
- Draw the circuit as in Figure 8.10, selecting TL071C as the operational amplifier.
- Set the component values as in the figure.
- Insert the two batteries.
- Click T&M followed by Oscilloscope and set the Time/Div to 1m, Volts/Div to 5.
- Click Run to observe the output waveform.

8.4.2 Square Wave Oscillator

Figure 8.11 shows a square wave output oscillator designed using an operational amplifier. The frequency of oscillation is given by:

$$f = \frac{1}{2RC \cdot \ln\left(\frac{2R1 + R2}{R2}\right)}$$

Where ln is the natural logarithm. If we assume R2 = 1.16 R1, then the frequency equation becomes:

$$f = \frac{1}{2RC \cdot \ln\left(\frac{3.16}{1.16}\right)}$$

ln(3.16/1.16) = 1, and therefore the frequency equation becomes:

$$f = \frac{1}{2RC}$$

The above equation can be used for designing the oscillator. In Figure 8.11, using the exact frequency equation, the frequency is given by:

$$f = \frac{1}{2 \cdot 10 \cdot 10^3 \cdot 0.05 \cdot 10^{-6} \cdot \ln\left(\frac{2 \cdot 10 \cdot 10^3 + 12 \cdot 10^3}{12 \cdot 10^3}\right)}$$

Which gives f = 1.019 kHz.

It is clear from Figure 8.11 that the frequency of oscillation is about 1 kHz.

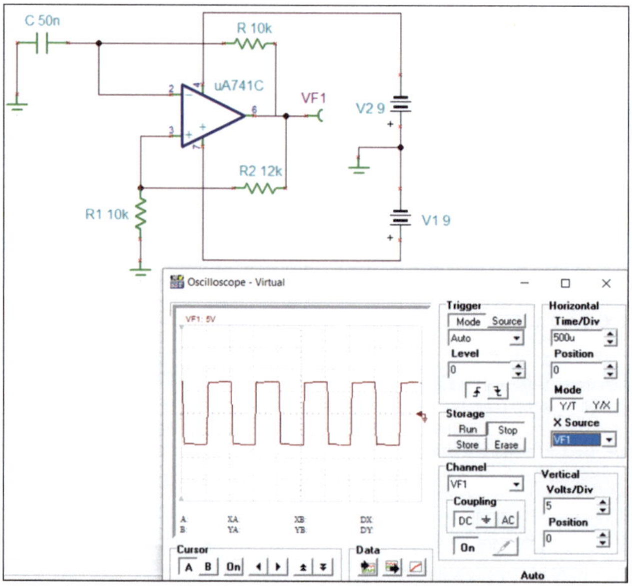

Figure 8.11 Square wave oscillator.

Procedure:

- Start the TINA-TI simulator.
- Draw the circuit as in Figure 8.11, selecting uA741C as the operational amplifier.
- Set the component values as in the figure.
- Insert the two batteries.
- Click T&M followed by Oscilloscope and set the Time/Div to 1m, Volts/Div to 5.
- Click Run to observe the output waveform.

8.5 Filters

Filters are electrical circuits that allow certain frequencies to pass through, but block others. i.e. they filter out or attenuate the frequencies we do not want. The band of frequencies which are passed is called the **passband** of the filter. The frequency at which the transition between passing and rejecting input frequencies occurs is called the **cut-off f**requency. The band of frequencies which are attenuated is called the **stopband** of the filter.

There are basically two types of filters as far as the type of component used: **passive** filters are constructed using only passive components such as resistors, capacitors, and inductors. Passive filters do not have any amplifying elements. These types of filters are usually used in high-frequency circuits. **Active** filters are constructed using transistors, operational amplifiers, resistors and capacitors. These filters are usually used in low- to medium-frequency circuits. Although in this chapter, we are interested only in active filters, passive filter circuits will also be given for demonstration purposes.

The most commonly used filter types are low-pass, high-pass, band-pass, and band-stop. A freely available computer program will be used to derive the component values of the filters. In addition, only Butterworth-type filters will be considered in this chapter.

8.5.1 Filter Roll-Off and Filter Order

The effectiveness of a filter is measured by its roll-off, which is the steepness of a transfer function with frequency. The order of a filter determines the steepness. The higher the order, the steeper is the transfer function. As the order of a filter is increased, so does its complexity and the number of components it uses.

The roll-off is measured as a function of logarithmic frequency. Commonly used units of roll-off are decibels per decade (dB/decade), or decibels per octave (dB/octave). A decade is a tenfold increase in frequency, and an octave is a twofold increase in frequency. For example, a first-order low-pass filter with a −20 dB/decade roll-off has a transfer function that drops by 20 dB if the frequency is multiplied by ten.

As an example, Figure 8.12 shows a first-order low-pass filter response. Here, the cut-off frequency is 10 kHz, i.e. the frequency where the gain drops by −3 dB compared to its passband value of 0 dB (at 0 dB, output level = input level). In the stopband, the gain falls −20 dB over each decade, i.e. at 10 kHz, 100 kHz, and so on. The slope of the curve in the stopband is −20 dB/decade. We can also say that the slope of the curve in the stopband is −6 dB/octave, i.e. the gain falls −6 dB at 2 kHz, 4 kHz, 8 kHz, and so on. Similarly, in a second-order low-pass filter, the stopband curve falls by −40 dB/decade (or −12 dB/octave).

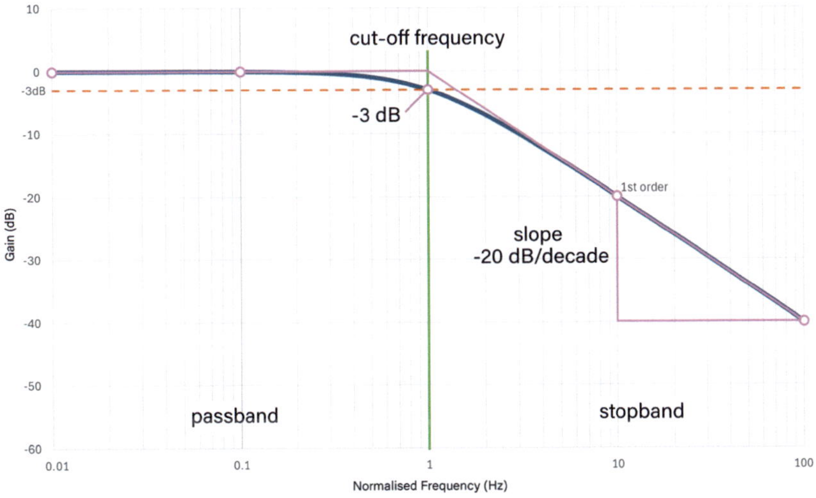

Figure 8.12 First-order −20 dB/decade low-pass filter response.

Figure 8.13 shows the change in the response curves for a low-pass filter as the filter order is increased.

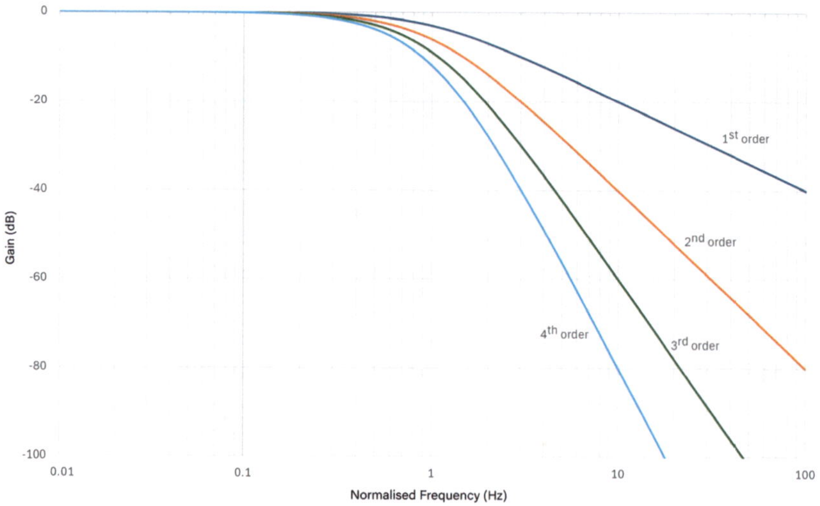

Figure 8.13 Frequency response and filter order.

8.5.2 Filter Design Program

The online Analog Filter Wizard design tool from Analog Devices is used in this chapter to calculate the component values and draw the circuit diagram of the filter to be built. The filter program is available online at the following site and there is no need to install it:

https://tools.analog.com/en/filterwizard/

The program can be used to design Butterworth and Chebyshev type filters. Only active filters are considered in this chapter.

8.5.3 Low-Pass Active Filters

A low-pass filter passes low frequencies up to its cut-off frequency and attenuates higher frequencies. At the cut-off frequency, the gain is down by -3 dB. At the stopband, the attenuation depends on the order of the filter used. An example low-pass filter design is given in this section.

Example

Design a low-pass active filter with the following specifications:

Passband gain: 0 dB (i.e. output level = input level)
Cut-off frequency = 10 kHz (i.e. −3 dB frequency)
Stopband slope: −40 dB/decade

Solution

The steps are:

- Start the filter design program.
- Set the passband gain to 0 dB, −3 dB point at 10 kHz and set the stopband slope to −39.9 dB at 100 kHz (i.e. the next decade. Notice that setting the value to −40 dB changes the slope to −60 dB). This design will correspond to a 2nd order filter.
- Make sure that the filter response shows Butterworth and click **Filter Response** (Figure 8.14).

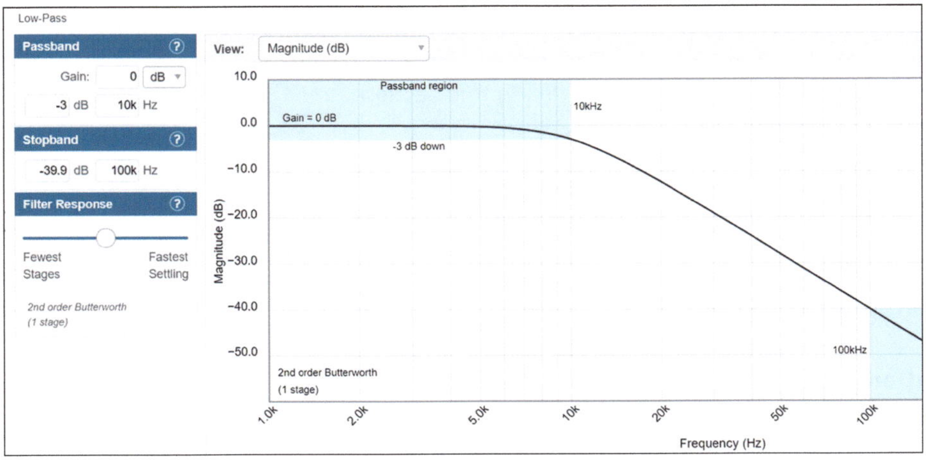

Figure 8.14 Second-order low-pass filter response.

- Click **Components** to display the circuit diagram of the filter (Figure 8.15). You can change the operational amplifier type, filter circuit type (Sallen & Key or Multiple Feedback), or the resistor and capacitor values.
- Click **Tolerances** and you can select the preferred capacitor and resistor series (e.g. E12, E24, etc.) so that you have standard-value components.

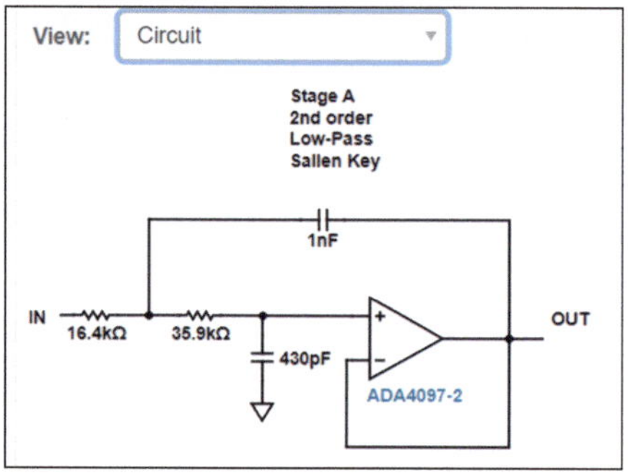

Figure 8.15 Second-order low-pass filter circuit.

8.5.4 High-Pass Active Filters

A high-pass filter passes frequencies above its cut-off frequency and attenuates frequencies below it. At the stopband, the attenuation depends on the order of the filter used. An example high-pass filter design is given in this section.

Example

Design a high-pass active filter with the following specifications:

Passband gain: 0 dB (i.e. output level = input level)
Cut-off frequency = 10 kHz (i.e. −3 dB frequency)
Stopband slope: −40 dB/decade

Solution

The steps are:

- Start the filter design program.
- Set the passband gain to 0 dB, −3 dB point at 10 kHz and set the stopband slope to −39.9 dB at 1 kHz (i.e. the previous decade. Notice that setting the value to −40 dB changes the slope to −60 dB). This design will correspond to a 2nd order filter.
- Make sure that the filter response shows Butterworth and click **Filter Response.**

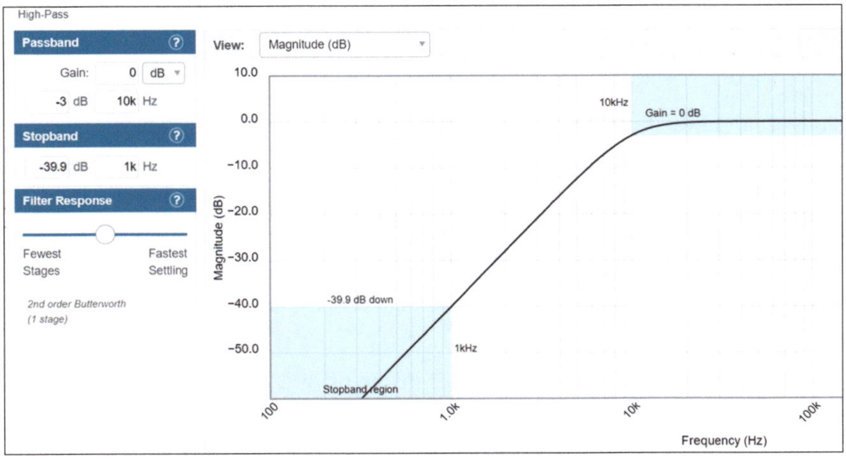

Figure 8.16 Second-order high-pass filter response.

- Click **Components** to display the circuit diagram of the filter (Figure 8.17). You can change the operational amplifier type, filter circuit type (Sallen & Key or Multiple Feedback), or the resistor and capacitor values.

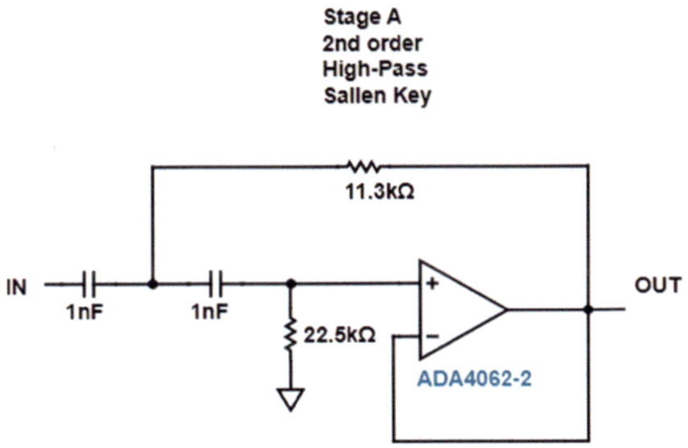

Figure 8.17 Second-order high-pass filter circuit.

8.5.5 Band-Pass Active Filters

A band-pass filter passes a band of frequencies centered around the center frequency and attenuates lower and higher frequencies outside this band. At the stopband, the attenuation depends on the order of the filter used. For the design of a band-pass filter, we have to specify the center frequency, the bandwidth (which is the width of the frequency band at the two −3 dB gain points), and the slope of the stopband curve. An example band-pass filter design is given in this section.

Example

Design a band-pass active filter with the following specifications:

> Passband gain: 0 dB (i.e. output level = input level)
> Bandwidth = 5 kHz
> Stopband curve = −40 dB at 60 kHz
> Center frequency = 10 kHz

Solution

- Start the filter design program.
- Set the passband gain to 0 dB, −3 dB point at 5 kHz (the bandwidth), set the stopband slope to −39.9 dB at 60 kHz (notice that setting the value to −40 dB changes the slope to −60 dB), and set the center frequency to 10 kHz.
- Make sure that the filter response shows Butterworth and click **Filter Response** (Figure 8.18).

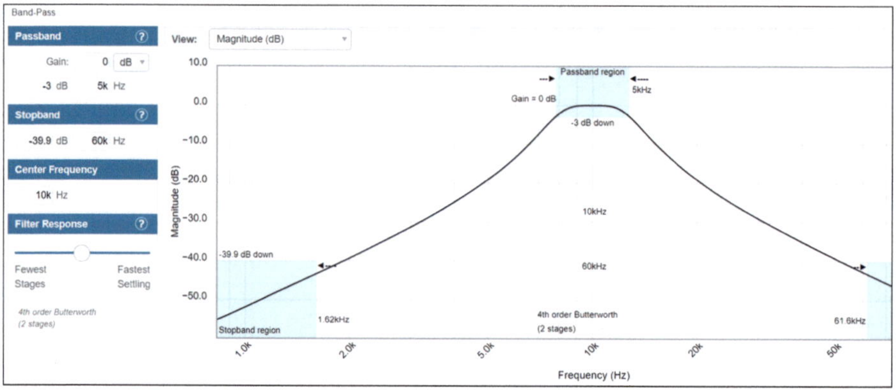

Figure 8.18 Fourth-order band-pass filter response.

- Click **Components** to display the circuit diagram of the filter (Figure 8.19).

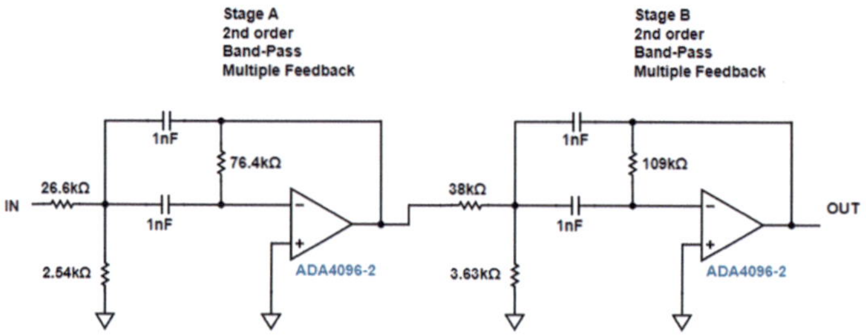

Figure 8.19 Fourth-order band-pass filter circuit.

8.5.6 Passive Low-Pass Filter

The passive low-pass filter shown in Figure 8.20 is commonly used in many electronic devices. This is a first-order filter where the cut-off frequency depends on the resistor and capacitor values. It is given by:

$$f_C = \frac{1}{2\pi RC}$$

There are several free online tools that can be used to calculate the resistor and capacitor values for the required cut-off frequency. Normally, the required cut-off frequency and a capacitor value are given. The program calculates the required resistor value. An example is given below.

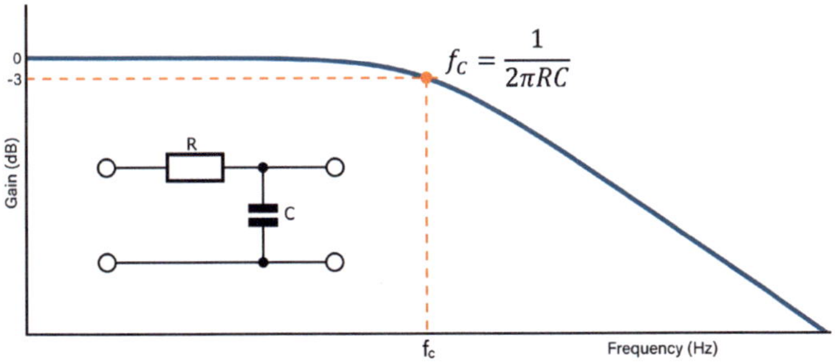

Figure 8.20 A first-order passive low-pass filter.

Example

Design a passive low-pass filter with a cut-off frequency of 1 kHz. Assume that the capacitor used is 1 μF.

Solution

In this example, the online design tool at the link below is used. There is no need to install this program.

https://www.learningaboutelectronics.com/Articles/Low-pass-filter-calculator.php

There are three parameters: the cut-off frequency, the resistor value, and the capacitor value. You should enter two of the parameters and the program will calculate and display the third parameter.

The steps are:

- Start the program.
- Enter the cut-off frequency as 1000 Hz and the capacitor value as 1 μF
- Click **Calculate**. The required resistor value is given as R = 159 Ω as shown in Figure 8.21.

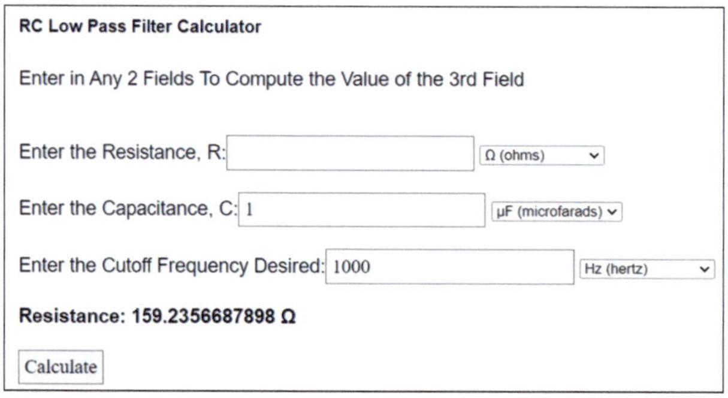

Figure 8.21 Calculate the required resistor value.

8.6 Sensors

A sensor is a device that responds to its physical environment. There are sensors for temperature, light, pressure, humidity, motion, force and so on. Sensors usually output electrical signals as current or voltage, depending on the measured value.

Sensors are used for automatic control, for monitoring, or for display purposes. For example, in a typical automatic room temperature control system, the room temperature is measured using a temperature sensor. Then, a microcontroller compares the measured temperature with the desired temperature and activates or deactivates a heater to keep the room temperature at the required level. In monitoring and display applications, the readings of a sensor are displayed on a visual display such as an LCD, LED, etc.

Sensors can be active or passive, depending on whether they require a power supply for their operation or not. The output of a sensor can be analog or digital. Digital output sensors can be directly connected to microcontroller inputs. The outputs of analog sensors, on the other hand, must be converted into digital using analog-to-digital converters (ADCs) before they can be connected to microcontroller systems.

Before looking at the specifications of some commonly used sensors, it is worthwhile to review the sensor terminology.

Range

The range of a sensor specifies the maximum and minimum values that can be measured. For example, a given temperature sensor may have a range of 0°C to 120°C. Such a sensor must not be used to measure temperatures outside these limits.

Resolution

The resolution of a sensor is the smallest detectable incremental change of input that can be measured by the sensor. For example, if the resolution of a temperature sensor is ±1°C, then this sensor cannot measure temperature changes smaller than 1°C.

Precision

This term refers to the reproducibility of a measurement using a sensor. In other words, if the same value is measured many times, one expects to get the same result every time.

However, a sensor may give slightly different results for each measurement, depending on its precision. For example, consider a force measuring sensor. If the force on the sensor is not changed, one would expect the sensor output to be the same at every reading. But in reality, small changes can occur between different readings.

Accuracy
Many people confuse accuracy with resolution or precision. The accuracy of a sensor is the maximum difference between the measured value and the actual value. For example, if the accuracy of a temperature sensor is 5°C, then we may expect that all its readings will be within ±5°C of the actual value.

Linearity
The linearity of a sensor is one of the important parameters. The output value of a linear sensor is directly proportional to the measured value. Its transfer function is a linear sloping line. Using linear sensors makes it straightforward to convert the measured value into a physical value.

Offset Error
The offset error of a sensor is its output value when its input is zero. For example, a force sensor may give a small output value even though there is no force exerted on it. The offset error is small in high-quality, expensive sensors.

Response Time
The output of a sensor does not change instantaneously when an input is applied. The time needed to produce a reading is called the response time of the sensor. For example, the output of a temperature sensor may delay, say by 100 ms when the temperature changes. Here, 100 ms is the response time of the sensor. In most applications, a small response time is desirable. If the response time is not respected by the device connected to the sensor, incorrect results will be obtained.

Power Supply
It is important to apply the correct power supply when using active sensors.

Self-Heating Effects
Some active sensors may self-heat while they are powered on, and this may change its output value slightly. For example, if a temperature sensor self-heats, then the measured temperature value will be in error. It is important to read the sensor specifications and take the necessary steps to avoid the self-heating problems. A sensor may also be heated by the system it is integrated into. This too may result in larger than expected measurement errors.

Hysteresis
The output of a sensor should follow the changes on its inputs, regardless of the direction of evolution (increasing or decreasing) of the measured parameter. This is known as the hysteresis property of the sensor. In high-quality sensors, the hysteresis effect is not desirable.

In the following section, we will look in more depth at some commonly used sensors.

8.6.1 Temperature Sensors

There are many types of temperature sensors. We will look at some of them and investigate their specifications.

Thermocouples

Thermocouples are simple temperature sensors consisting of two different metals. In 1821, German physicist Thomas Seebeck discovered that a voltage is produced and an electric current flows in a closed circuit of two different metals if the two sides of their junction are held at different temperatures.

There are about 12 standard thermocouple types that are commonly used. Each type is given an internationally approved letter that indicates the materials that the thermocouple is manufactured from. For example, the Type K thermocouple is constructed using Ni-Cr and Ni-Al metals. It is low-cost and one of the most popular thermocouples. The operating range is from −180°C to +1,350°C with a very high sensitivity. This type of thermocouple is most suited to oxidizing environments. Some other types of thermocouples are: Type J, Type N, Type T, Type E, Type B, and so on. Thermocouple sensors are available as hand-held probes, as bolts, washers, and terminal heads. Figure 8.22 shows some thermocouples.

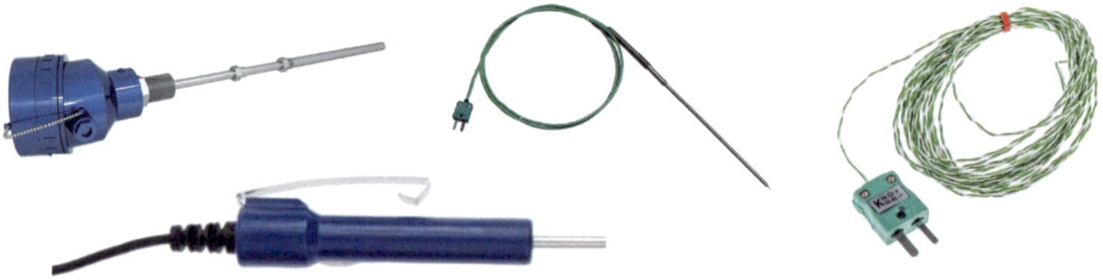

Figure 8.22 A few thermocouples.

Resistance Temperature Detector (RTD)

The RTD is a temperature sensing device whose resistance increases with temperature. RTDs are very linear devices, and they have excellent accuracy over a wide temperature range. Some RTDs have accuracy better than 0.001°C. Another advantage is that their values do not drift with ageing.

Every piece of metal has a resistance that is proportional to the length and the resistivity of the piece of metal, and inversely proportional to its cross-sectional area:

$$R=\frac{\rho L}{A}$$

Here, R is the resistance in ohms, ρ is the resistivity (in ohm-meters) which increases with increasing temperature, L is the length of the metal (in m), and A is the cross-sectional area (in m^2).

The most used international RTD standard is IEC 751. Other standards such as BS 1904 and DIN 43760 match this standard. The IEC 751 standard is based on platinum RTDs with a resistance of 100 Ω at 0°C.

RTDs are used in resistive potential divider circuits with a known resistor. By applying an input voltage and measuring the output voltage, we can calculate the resistance and hence the temperature of the RTD.

Thermistors

The name thermistor derives from the words 'thermal' and 'resistor'. Thermistors are temperature sensitive passive semiconductors which exhibit a change in electrical resistance when subjected to a small change in body temperature. As shown in Figure 8.23, thermistors are manufactured in various sizes and shapes.

Figure 8.23 Thermistors

There are two types of thermistors: PTC and NTC. NTC stands for Negative Temperature Coefficient; PTC means Positive Temperature Coefficient. Negative means that the sensor's electrical resistance decreases as its temperature increases. Positive is the opposite. The relationship between the resistance and temperature is nonlinear.

NTCs are used in low-cost temperature measurement applications. The advantages of NTCs are:

Sensitivity: One of the advantages of thermistors compared with thermocouples and RTDs is their relatively large change in resistance with temperature.

Small size: Thermistors are small, and this makes for a very rapid response to temperature changes.

Ruggedness: Most thermistors are rugged and can handle mechanical and thermal shock and vibration better than other types of temperature sensors.

Remote measurement: Thermistors can be used to sense the temperature of remote locations via long cables.

Low cost: Thermistors cost less than most other types of temperature sensors.

The following equation is commonly used to calculate the resistance of a thermistor at a given temperature:

$$R_T = R_{25} \cdot e^{\left(\frac{\beta}{T_T} - \frac{\beta}{T_{25}}\right)}$$

Where R_T is the resistance of the thermistor, R_{25} is the resistance at 25°C, T_{25} is 25°C expressed in Kelvin (i.e. 298.15 K), β is the temperature constant of the thermistor, and T_T is the temperature of the thermistor in Kelvin.

However, what we want is to measure the resistance R_T and then calculate the temperature T_T of the thermistor. This is given by:

$$T_T = \frac{1}{\frac{1}{\beta}\ln\left(\frac{R_T}{R_{25}}\right) + \frac{1}{T_{25}}}$$

Where ln is the natural logarithm. An example is given below.

Example

The temperature constant of a thermistor is β = 2910. Also, the resistance of the thermistor at 25°C is R_{25} = 1 kΩ. This thermistor is used in a circuit to measure the temperature. The resistance of the thermistor is measured as 800 Ω. Calculate the temperature.

Solution

First, the temperature at 25°C is converted into K:

$$T_{25} = 25 + 273.15 = 298.15 \text{ K}$$

Using the above equation:

$$T_T = \frac{1}{\frac{1}{2910}\ln\left(\frac{800}{1000}\right) + \frac{1}{298.15}}$$

Which gives T_T = 305.13 K, or, in Celsius, T_T = 305.13 – 273.15 = 31.98 = 32°C.

Integrated Circuit Temperature Sensors

IC-based temperature sensors are semiconductor devices fabricated similar to other semiconductor devices. IC temperature sensors differ from other sensors in some fundamental ways:

- These sensors have relatively small sizes.
- Their outputs are linear.
- The temperature range is limited.

- The cost is relatively low.
- These sensors can include advanced features such as thermostat functions, built in analog-to-digital converters, alarm circuits and so on.
- Often these sensors do not have good thermal contacts with the outside world and as a result, it is more difficult to use them other than in measuring the air temperature.
- A power supply is required to operate them, i.e. they are active sensors.

IC temperature sensors can be divided into the following categories:

A. Analog-output sensors
B. Digital-output sensors

Analog output sensors are further divided into:

A. Voltage output sensors
B. Current output sensors

Voltage-Output Temperature Sensors
These sensors provide a voltage output which is proportional to the temperature measured. Some popular sensors in this group are:

LM35D: This is a 3-pin sensor IC (Figure 8.24) with a range from 0°C to +100°C, with a maximum error of ±1°C. The output voltage is given by:

$$V_O = 10 \text{ mV per degree Celsius}$$

For example, at 10°C, the output voltage is 100 mV; at 30°C, the output voltage is 300 mV, and so on.

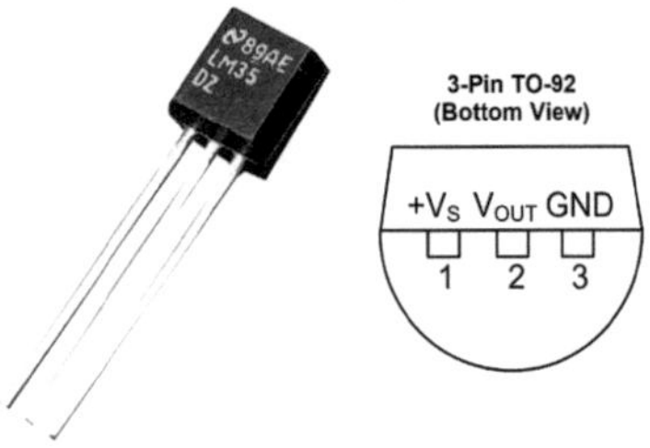

Figure 8.24 The LM35 temperature sensor.

There are several variants of the LM35. For example, LM35C has a range of −40°C to +110°C. The **LM34** is used to measure temperature in °F. The **LM50** can measure negative temperatures without any external components.

Current-Output Temperature Sensors
These sensors provide a current output which is proportional to the temperature measured. Some popular sensors in this group are:

AD590: This sensor (Figure8.25) has a range from −55°C to +150°C and the output current is proportional to measured voltage given by:

I = 1 μA per Kelvin

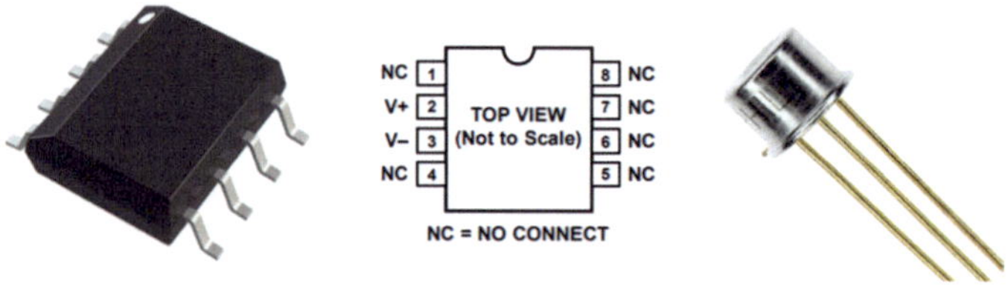

Figure 8.25 The AD590 temperature sensor.

The **LM134** is a programmable sensor with an output current of 0.1 μA/K to 4 μA/K. The temperature range of this sensor is −25°C to +100°C. The LM134 typically needs only a 1.2 V supply voltage to operate.

Digital-Output Temperature Sensors
These sensors produce digital outputs which can be interfaced directly to a computer-based system, e.g. a microcontroller. The outputs are usually non-standard and the temperature reading can be extracted by using suitable algorithms. Some examples of digital-output sensors are:

LM75: This sensor (Figure 8.26) operates within the temperature range from −55°C to +125°C and provides I²C-compatible 9-bit serial output with a resolution of 0.5°C. The device can be programmed to monitor the temperature. It can also set an output high or low if the temperature exceeds a preprogrammed value.

The **DS18B20** is a 1-wire digital temperature sensor, meaning it communicates over a single data wire. It has a wide temperature measurement range, from −55°C to +125°C and is known for its high accuracy. Each DS18B20 sensor has a unique 64-bit address, allowing multiple sensors to be connected to the same 1-wire bus. The DS18B20 can be powered in a 'parasitic' mode, where it draws power from the data line during temperature conversion, simplifying wiring.

Some other digital-output temperature sensors are: TMP03, AD7814, MAX6575, MAX31855, TMP422 and so on.

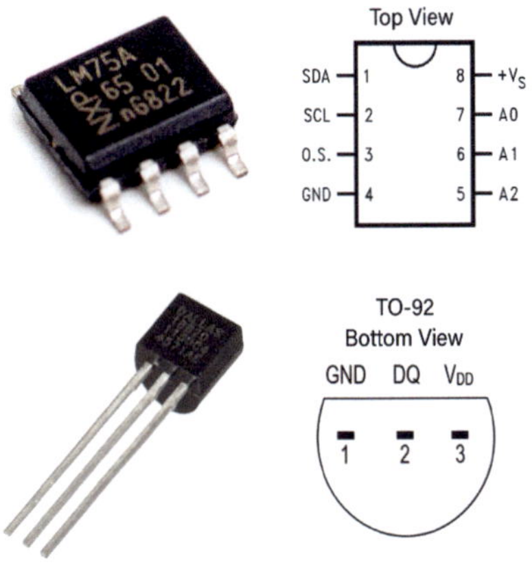

Figure 8.26 The LM75 (top) and the DS18B20 (bottom).

One of the recently popular temperature and humidity sensor ICs are the **DHT11** and **DHT22** (Figure 8.27). These chips operate from 3 V to 5 V and provide a digital output. The DHT11 is the cheaper version, and its range is from 0°C to +50°C for temperature and from 20% to 80% for humidity measurements. The DHT22 has a wider range, from −40°C to +80°C for temperature and from 0% to 100% for humidity.

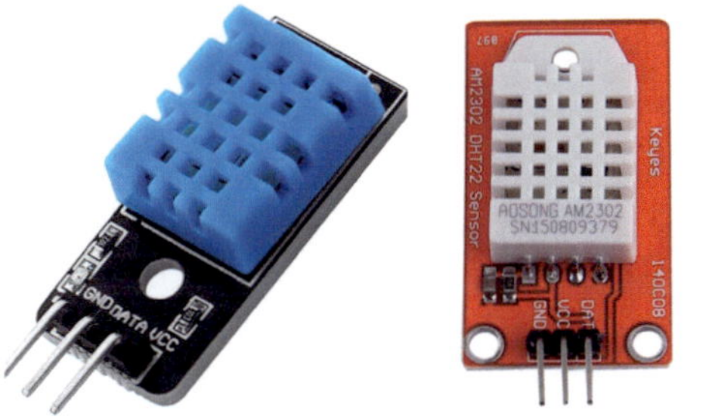

Figure 8.27 The DHT11 (left) and DHT22 (right).

8.6.2 Force Sensors

Force sensors transform inputs such as force, weight, torque, compression, stress, or strain into electrical signals. Almost all force sensors are active, requiring a power supply for their operation. The outputs of force sensors can be analog or digital. Force sensors can be in the form of load cells used in weighing platforms and in beams with capacities ranging to up to 50,000 kg. Figure 8.28a shows a button-type load cell used in compression measurements. Figure 8.28b shows a force-sensitive resistive film, which is used to measure loads up to 30 kg while having a diameter of only 30 mm. The resistance of this sensor changes with the applied force. Figure 8.28c shows a piezoresistive force sensor offering analog or digital outputs with a 3.3 V supply voltage. The load range of this sensor can be up to 1.2 kg. Figure 8.28d shows a piezoelectric force sensor that is used to measure force in tension or compression. This sensor includes an internal amplifier for high sensitivity.

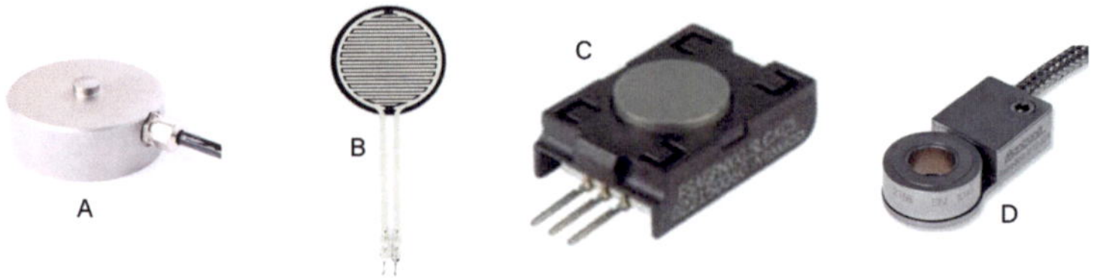

Figure 8.28 A selection of force sensors.

8.6.3 Light Sensors

The most commonly used light sensor is the Light-Dependent Resistor (LDR) as was discussed in earlier chapters. This is basically a resistor whose resistance is very high in the dark and very low when light falls on the device (Figure 8.29).

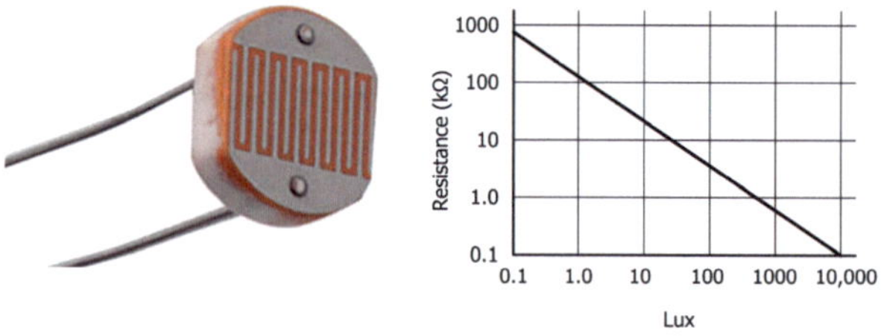

Figure 8.29 Light-dependent resistor.

8.6.4 Sound Sensors

A sound sensor detects sound waves and outputs electrical signals depending on the intensity of the sound. They function like ears, which have a vibrating diaphragm. Perhaps the commonest sound sensor is the microphone that everyone is familiar with. Sound sensors are used in many applications, such as in music systems, computers, burglar alarms, ambient sound level recognition, home automation, etc.

There are basically three types of microphones: dynamic, condenser, and ribbon. A dynamic microphone has a moving coil and a diaphragm. It is used when sound levels are high. Dynamic microphones are low-cost.

Condenser microphones have a thin conductive diaphragm close to a metal backplate, which works like a capacitor. The changes in the capacitance are converted into electrical output signals. Because there is no coil, condenser microphones have better sound quality. MEMS microphones are often condenser types.

Ribbon microphones are commonly used in the radio industry. They work on the same principle as dynamic microphones, but they use a strip of corrugated aluminium instead of a moving coil. Ribbon microphones capture natural sound compared to other types of microphones. They are more sensitive, especially on higher frequencies.

Sound level is measured in decibels (dB). If the sound decibel level increases by 10, the sound intensity is multiplied by 10. Because the decibel is a logarithm entity, a 10 dB sound is 10 times more intense than a 0 dB sound. Normal level conversation is about 50–60 dB, while, for example, the sound level of a jet taking off is about 120 dB. Sound levels above 120 dB are dangerous for the human ear, as they can cause irreversible damage, including deafness.

8.6.5 Acceleration Sensors

Acceleration sensors or accelerometers are used to measure acceleration, vibration, shock, and movement in various applications. Most accelerometers are tiny devices based on MEMS (Micro-Electro-Mechanical System) technology. Some accelerometers operate on two axes (2D), while more advanced ones can operate on three axes (3D).

An accelerometer measures g, acceleration due to gravity. The MPU6050 (Figure 8.30a) is a combined 3-axis accelerometer and 3-axis gyroscope integrated on a single chip and is widely used in microcontroller-based systems. The accelerometer can measure gravitational acceleration along the three axes. The output of the gyroscope is in degrees per second.

Another commonly used accelerometer is the ADXL345 (Figure 8.30b). It is a popular 3-axis accelerometer with the following specifications:

- 2.0 V to 3.6 V operating voltage
- Ultra-small current consumption (23 μA)
- 13-bit resolution at ±16 g
- Free fall detection
- 10,000 g shock survival
- SPI and I²C interface
- Wide temperature range

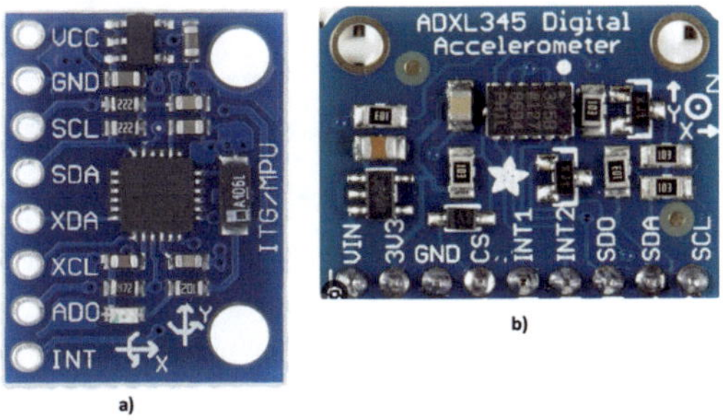

Figure 8.30 The MPU6050 and ADXL345 accelerometers on ready-to-use modules.

8.6.6 Soil Moisture Sensors

A soil moisture sensor is a hygrometer used for detecting soil moisture levels. Figure 8.31 shows a commonly used sensor which provides both analog and digital outputs and is available as a module. Operating from 3.3 V to 5 V, the sensitivity is adjustable.

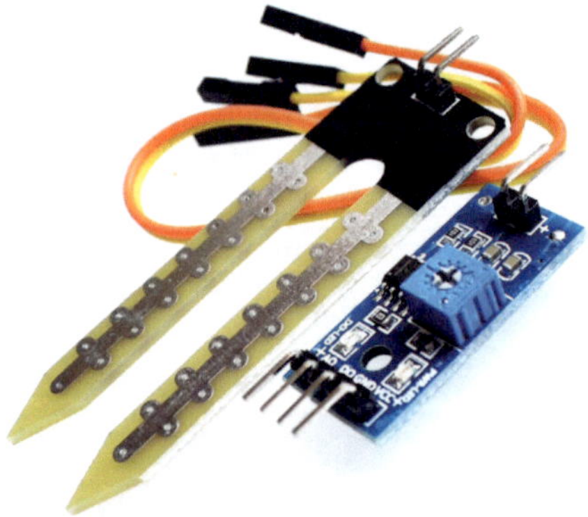

Figure 8.31 Soil moisture sensor with signal processing module.

8.6.7 Distance Sensors

A Time-of-Flight (ToF) sensor is a type of sensor that measures the time it takes for an object, light, or sound wave to travel a certain distance. Time-of-Flight sensors are commonly used for distance or depth sensing and can be based on various principles, including light (lidar, IR), sound and ultrasound, or radio waves (radar).

The HC-SR04 (Figure 8.32) is a popular ultrasonic ToF sensor providing 2 cm to 400 cm distance measurement with an accuracy of up to 3 mm. The transmitter ('T') sends pulses, which are detected by the receiver ('R'). The time taken by the reflected pulses to return to the receiver is used to determine the distances to the objects in front of the sensor.

Figure 8.32 HC-SR04 ultrasonic transducer.

8.7 Switched-Mode Power Supply (SMPS)

A Switched-Mode Power Supply (SMPS) is a type of power supply that uses a switching regulator to efficiently convert electrical power. It operates by rapidly switching the input voltage on and off at a high frequency, typically in the kilohertz or megahertz range.

When an SMPS converts an input voltage to a lower output voltage, it is called a buck or step-down converter. When it converts an input voltage to a higher output voltage, it is called a boost or step-up converter.

Buck and boost converters are very efficient devices, with up to 90% efficiency or more. High efficiency extends battery life, reduces heat dissipation, and allows for smaller and more portable electronic devices to be built. Buck converters are commonly used in computers to lower the 12 V or 15 V DC supply down to 5 V, 3.3 V, or 1.8 V. Boost converters, on the other hand, are used to increase a voltage from e.g. a 1.5 V battery to 3.3 V.

Buck and boost converters are usually based on a switching transistor, a diode, a capacitor, and an inductor. In this section, we will be looking at both types of these DC-to-DC converters: a buck converter with an LM2576 and a boost converter with an LM2577.

8.7.1 The LM2576 Buck Converter

The LM2576 buck converter is a step-down switching regulator with a maximum input voltage of 40 V, capable of driving a 3 A load. It is available with fixed-output voltages (3.3 V, 5.0 V, 12 V, and 15 V) and with an adjustable output. It has an efficiency of up to 90%. The output voltage is mentioned in the name, such as LM2576-3.3 (3.3 V), LM2576-5.0 (5 V), etc. or LM2576-ADJ for the adjustable output version. The device includes thermal and current limit protection and requires only a few external components for its operation. For example, Figure 8.33 shows a circuit capable of providing +5 V, 3 A output from a 7 V to 40 V unregulated DC input voltage.

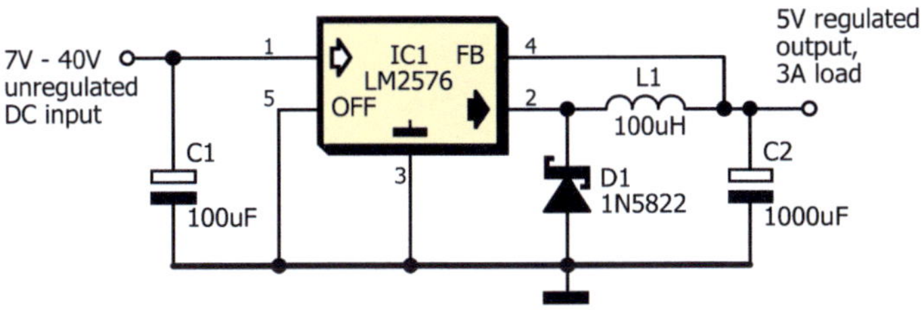

Figure 8.33 LM2576-based buck converter example circuit.

8.7.2 The LM2577 Boost Converter

The LM2577 is a boost converter with an input voltage range from 3.5 V to 40 V, the output voltage can go up to 60 V. It can deliver up to 3 A of current. Available output voltages are: 12 V (LM2577-12), 15 V (LM2577-15), and adjustable (LM2577-ADJ). The output is protected by a current limiter and over temperature protection. Only a few external components are required to build a boost converter. Figure 8.34 shows an example circuit with an output voltage that can be set with the two resistors R1 and R2. Note that the input voltage must always be at least 2 V lower than the output voltage.

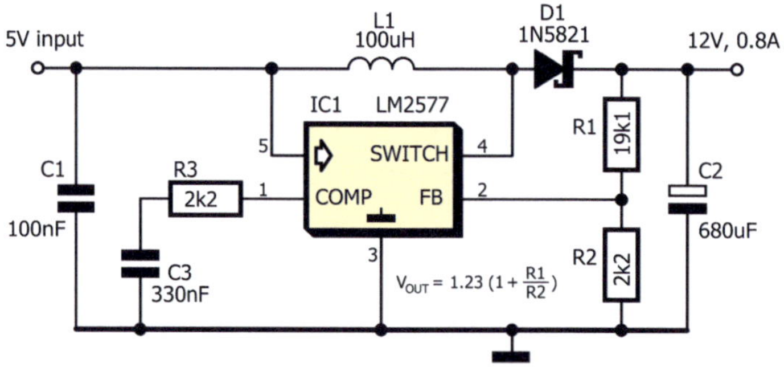

Figure 8.34 LM2577-based boost converter example circuit.

8.8 Interesting Electronic Kits

There are many interesting electronic kits on the Internet that you may like to purchase and experiment with. Some of these are described in this section.

Velleman Kits

Velleman (velleman.co.uk) offers many interesting kits that you may like to build, test, and use. For example:

- 3 – 30 V, 3 A power supply kit (Figure 8.35a)
- Component tester kit (Figure 8.35b)
- Temperature sensor kit (Figure 8.35c)
- Electronic dice kit (Figure 8.35d)
- Code lock kit (Figure 8.35e)
- Timer kit (Figure 8.35f)
- And many others

Figure 8.35 Some interesting Velleman kits.

Kitronik kits

Kitronik (kitronik.co.uk) offers some interesting electronic kits:

- Mono amplifier kit (Figure 8.36a)
- FM radio kit (Figure 8.36b)
- Steady hand game kit (Figure 8.36c)
- Solar power kit (Figure 8.36d)
- And many others

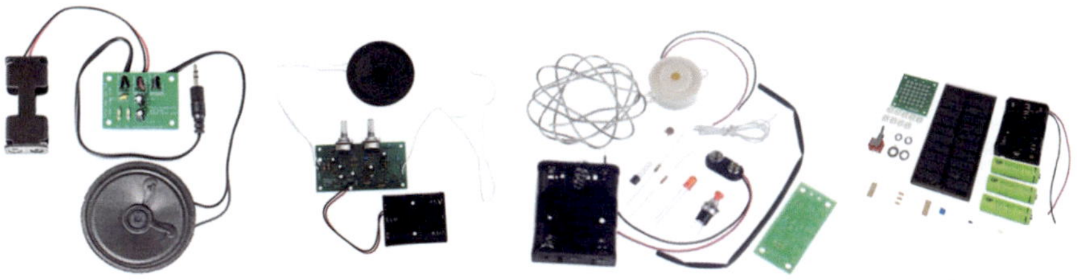

Figure 8.36 Some interesting Kitronik kits.

Microsoft Open Source

Many easy-to-build simple cardboard circuit kits are offered by Microsoft Open Source (https://microsoft.github.io/cardboard-circuits/). Some examples are a game controller, a race car, on-off projects, a switch, a button, LED projects, and so on.

Quasar Electronics Kits

Quasar electronics (https://quasarelectronics.co.uk/Category/kits-modules-educational-fun-projects) offers many educational and fun kits that you may like to build. Some projects are model trains, radios, robots, sound devices, solar kits, test and measurement tools, games, and many more.

Chapter 9 • Test and Measurement

9.1 Overview

Electronic test and measurement equipment are used to create various signals, to capture responses from electronic circuits, and to measure various parameters such as voltages and currents in circuits.

Many different types of electronic test and measuring equipment are required for designing and analysing all kinds of electronic circuits. In most general-purpose applications, the following basic equipment should be enough:

- Power supply
- Multimeter
- Oscilloscope
- Test probes
- Function generator

For specialized high-frequency and radio electronics work, the following additional equipment may be useful:

- RF signal generator
- RF probe
- Signal tracer
- Frequency counter
- Capacitance and inductance meter
- Spectrum analyser

In this chapter, we will be looking at the specifications of some commonly used basic equipment.

9.2 Power Supplies

Power supplies provide DC output voltages, and they come in many different shapes and sizes. They are always required in electronics laboratories. The important points to consider before purchasing a power supply are:

9.2.1 Fixed or Variable Voltage

Some power supplies provide only fixed voltages, e.g. 5 V or 12 V. These power supplies are rather limited since they cannot be used in applications that may require for example 9 V or 24 V or other voltages. It is best to purchase a power supply with variable output voltage. Usually, an analog meter or a digital display is provided on the front of the device to adjust the required output voltage through a knob.

9.2.2 Number of Output Channels

Single-channel power supplies provide only one output. Dual-channel or triple-channel power supplies provide more than one independent output voltage. In many situations, it is practical to have at least a dual-channel power supply.

9.2.3 Current Capacity

This is another essential specification of a power supply. In most microcontroller-based electronic projects, current capacities of up to a few amperes should be sufficient. As with the voltage, an analog meter or a digital display is provided to show the maximum current or the actual current at any time.

Figure 9.1 shows a dual-output power supply with digital displays for each channel, showing the voltage and maximum current for each channel.

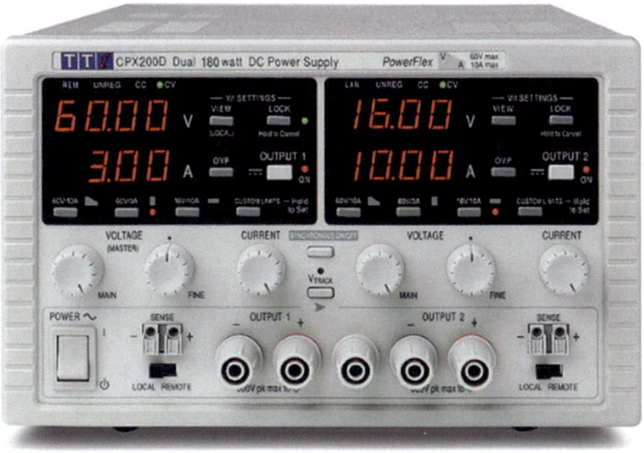

Figure 9.1 Dual-output power supply (source: www.aimtti.com).

9.2.4 Current Limiting

Most commercial power supplies feature current limiting. This serves to protect the connected circuit against strong currents due to short circuits, manipulation errors, etc. When the current limit is exceeded, the power supply shuts down its output automatically to switch off the connected circuit. The actual current drawn from the power supply at any time is usually shown on a display on the power supply.

9.3 Multimeter

A multimeter is the most important equipment to have while working with electronic circuits. Today, almost all multimeters are digital and have LCD-type displays. There are hundreds of different types of multimeters available on the market, with prices ranging from tens of euros to hundreds of euros. Depending on the type of multimeter, the following parameters can be measured:

- AC and DC voltage
- AC and DC current
- Resistance
- Continuity
- Capacitance
- Inductance
- Frequency
- Temperature
- Transistors and diodes

Additionally, external components can be attached to some multimeters to measure:

- Light intensity
- Sound pressure
- pH
- Humidity
- Very small voltages and currents

All multimeters can be used to measure voltage, current, resistance, and continuity. Figure 9.2, left, shows a typical low-cost multimeter costing around €15 that is good enough for most of the measurements you may need to do. Of course, you can spend much more money for a multimeter having more functions and even an oscilloscope-type display (Figure 9.2, right).

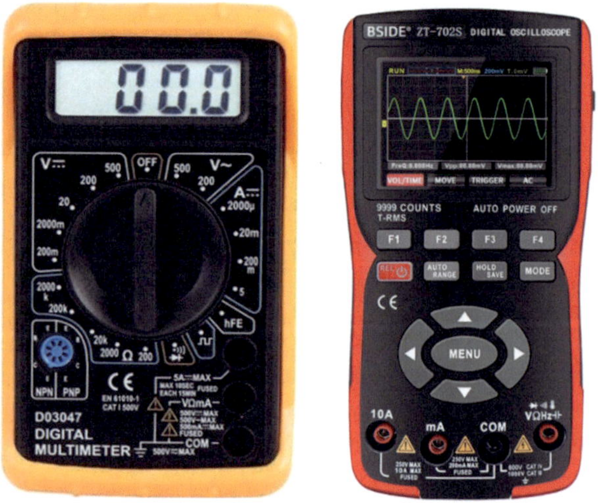

Figure 9.2 Low-cost multimeter (left, source: Duratool) and an oscilloscope/multimeter (right, source: zoyitools.com).

9.3.1 Parts of a Multimeter

Let us look at closely at the basic multimeter shown in Figure 9.2. Figure 9.3 shows various parts of this multimeter. This particular multimeter has the following specifications:

- 3½-digit display
- DC voltage: 200 mV to 500 V
- AC voltage: 200 V and 500 V
- DC current: 2,000 μA (i.e. 2 mA) to 5A
- Resistance: 200 Ω to 2,000 kΩ (i.e. 2 MΩ)
- Diode and continuity tester
- Square wave output
- NPN, PNP transistor tester
- Transistor hFE (β) reading

- Common and V/Ω/mA probe socket
- High current (5A) probe socket
- 9 V PP-3 battery operation

Figure 9.3 Parts of a multimeter.

9.3.2 Voltage Measurement

- Connect the black test probe to the COM socket and the red probe to the V/Ω/mA socket.

- If you will be measuring DC voltage, rotate the switch to the DC voltage region. If, on the other hand, you will be measuring AC voltage, rotate the switch to the AC voltage region. If you know the maximum value of the voltage to be measured, then select this on the voltage region. If you are not sure what the maximum voltage is, select the highest value first, and then you can rotate the switch to lower values if needed. In most low-voltage electronic circuits, the 20 V range should be OK.

- Put the red test probe on the positive side of the circuit and the black probe on the negative side of the circuit (i.e. across the part of the circuit of which you want to measure the voltage). It does not matter if you connect them the wrong way around (e.g. if the reading is negative but the voltage you are trying to measure is positive), just reverse them.

- The voltage reading will be shown on the display.

Figure 9.4 shows how the voltage across a resistor can be measured.

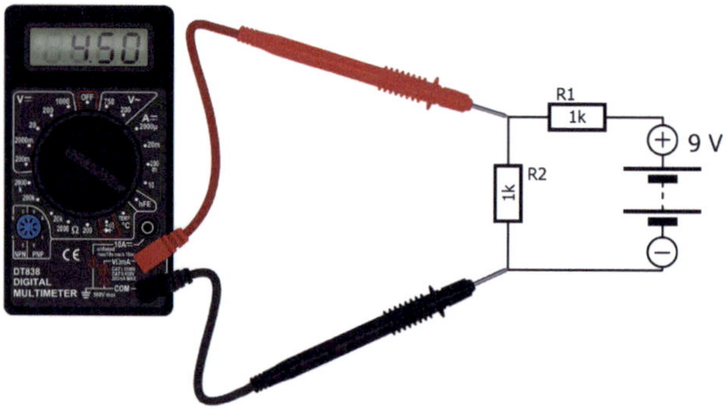

Figure 9.4 Measuring the voltage across a resistor. Note the position of the range selector switch.

Note: Some multimeters have an auto-ranging setting such that the voltage, current, resistance, etc. ranges will be selected automatically. If you have an auto-ranging multimeter, all you have to do is turn the switch to the correct parameter to be measured (e.g. voltage, current, resistance, etc.) and the range will be selected for you automatically.

9.3.3 Current Measurement

Measuring a current is similar to measuring a voltage, but the test probes must be connected in series with the component through which the current to be measured flows. Also, make sure you select the current region on the multimeter. Figure 9.5 shows how the current through a resistor can be measured.

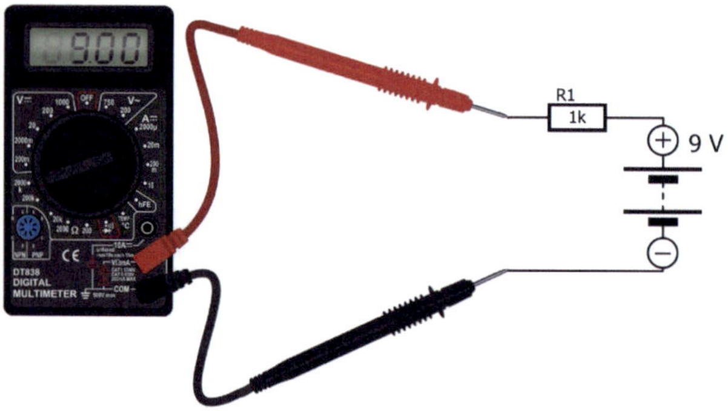

Figure 9.5 Measuring current through a resistor. Note the position of the range selector switch.

Warning: Make sure that the current you want to measure is within the maximum current rating of the multimeter. Unless the multimeter is auto-ranging, you should always select the maximum range for current measurements and then decrease the range for more accurate results. If the meter has a special probe socket for high currents, use that one first. A good multimeter is equipped with a fast-acting fuse to protect the instrument when more than the rated maximum current is flowing through it, but these fuses can be expensive to replace.

9.3.4 Resistance Measurement

Resistance measurement is similar to voltage measurement. Make sure that the resistor is not in a circuit (Figure 9.6). Set the multimeter to the resistor region and connect the resistor terminals to the probes of the multimeter. The value of the resistor will be displayed on the LCD. You may have to change the resistor range unless your multimeter can automatically set itself to the best range.

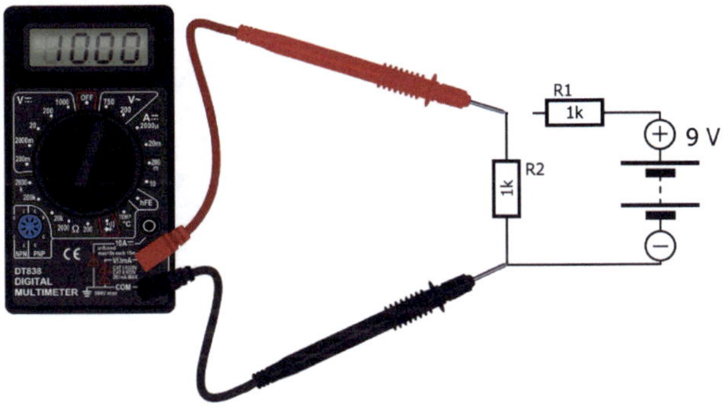

Figure 9.6 Measuring resistance. Note the position of the range selector switch.

Keep in mind that your body has resistance too. Therefore, if you clamp both probe tips with your fingers against the leads of a component, you may not always obtain a correct reading. Always try to keep at least one probe tip isolated from your body.

9.4 The Oscilloscope

An oscilloscope is a good investment as it is a very practical and versatile instrument. An oscilloscope displays graphically voltages as a function of time. The displayed waveforms can be analysed and its important parameters such as the frequency, amplitude, and shape can easily be determined. Most oscilloscopes are nowadays digital, and they are small and cheap compared to older, analog oscilloscopes. Furthermore, many of them can connect to a PC, allowing waveforms to be recorded and displayed on the PC screen. The PC also lets you control some oscilloscopes remotely.

A typical digital oscilloscope is shown in Figure 9.7. This is a 2-channel oscilloscope which has two inputs, and therefore it can display two waveforms at the same time.

The screen of an oscilloscope has a grid overlaid on it. If Volts/Div is set to 2, then each vertical grid division (i.e. each horizontal grid line) corresponds to 2 V. This way, you can measure the amplitude of a waveform displayed on the screen.

The horizontal time axis is controlled by the Time/Div region. For example, if Time/Div is set to 2 ms, then each horizontal grid division (i.e. each vertical grid line) is 2 ms. This way, you can measure the period of a waveform displayed on the screen.

The vertical position of a waveform on the screen can be moved up or down using the Y-pos buttons. Similarly, the horizontal position of the waveforms on the screen can be shifted using the H-pos button. The center of the screen is where voltage and time are zero, unless you turned a V-pos or the H-pos button. Most oscilloscopes have a lot more settings than mentioned here.

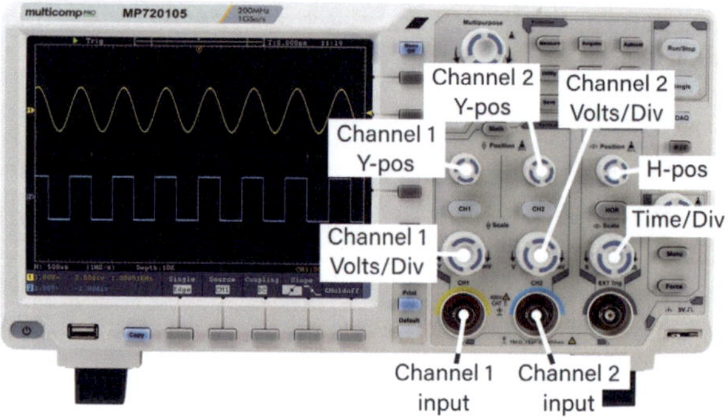

Figure 9.7 A typical dual-channel oscilloscope (source: uk.farnell.com).

9.4.1 Specifications

When selecting an oscilloscope, there are many different specifications to consider, and it is important to know what they are.

Bandwidth

Perhaps the most important specification of an oscilloscope is its bandwidth. In general, oscilloscopes with higher bandwidths cost more. High-bandwidth oscilloscopes can be used in high-frequency applications. For example, if you have a 50-MHz oscilloscope, you should be able to measure frequencies up to 50 MHz. Above that frequency, the oscilloscope might still be usable, but the waveform may no longer reflect reality.

Sample Rate

The sample rate is the number of samples that the oscilloscope can capture every second. The higher the sample rate, the better it is. However, high sample rates require more memory and faster processors and therefore make oscilloscopes more expensive.

Memory Size

The memory size (or the memory depth) is related to the sample rate. As the oscilloscope samples the waveforms, it stores them in memory. For example, an oscilloscope having a memory size of 50 MB can record up to 50 million samples in its memory. If there are two channels, then this number must be divided by 2.

Number of Channels

Most oscilloscopes have more than one channel. Two channels are enough for almost all applications that you might encounter. The timing of both channels is the same. If two channels are not enough, there are 4-channel and 8-channel oscilloscopes, but it may be more economic to invest in two 2-channel instruments.

Vertical Channel Resolution

Digital oscilloscopes convert the analog input signal into a digital signal. The vertical channel resolution determines the accuracy of the vertical voltage scale. Most low-cost oscilloscopes have 8-bit resolution, corresponding to 256 digitizing levels. In a scope with 8-bits of vertical resolution, only 256 levels are available for drawing the vertical scale. More expensive oscilloscopes have 10- or 12-bit resolution, making them more accurate to draw the waveforms.

The Rise Time

This is another important specification when it is required to capture fast rising signals. The rise time of an oscilloscope is defined as the time it takes for the image to rise from 10% to 90% of its final value.

Screen Size

A large screen is comfortable and allows for showing more information. The cost of an oscilloscope goes up as the screen size is increased. The screen size of PC-based oscilloscopes could be the same size as the PC screen.

Time Base

The time base defines the time resolution of the oscilloscope. Inspecting for instance the rise time of a pulse requires a high resolution. On low-cost oscilloscopes, the highest resolution is 10 ns/div or better. People often tend to only consider the fastest specifications, yet an oscilloscope with a time base that extends up to, say 1000 s/div can be very interesting as it allows capturing slowly changing signals that are hard to interpret otherwise.

9.4.2 Using the Oscilloscope

The oscilloscope is a complicated instrument, and it takes time and practice to learn to use it. Reading the manual is a good way to get started. If you are new to the oscilloscope, then you may find it difficult to measure the period, frequency, and the amplitude of a waveform. An example is given in this section to show how the period and amplitude can be read, and how the frequency can be calculated.

Example

Figure 9.8 shows a sine wave captured on an oscilloscope. In this figure, the horizontal scale was set to 500 μs/div and the vertical scale was set to 1.0 V/div. Determine the period and amplitude of this waveform, and then calculate its frequency.

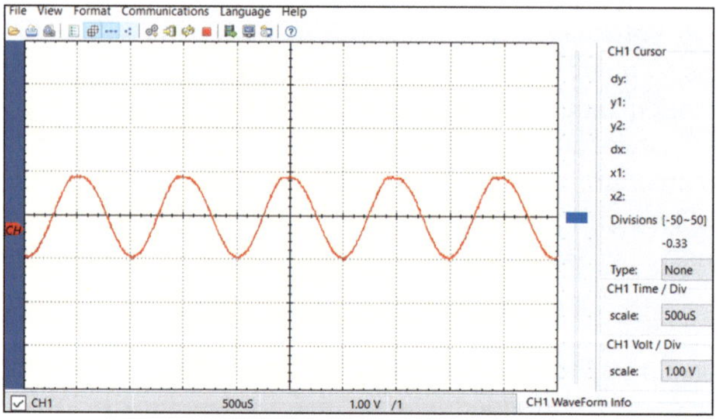

Figure 9.8 Sample sine wave.

Solution

Looking at the figure, we see that one period of the waveform needs two (horizontal) divisions. With the time base set to 500 μs/div, this is equal to 1000 μs. Therefore, the period of the waveform T = 1000 μs or T = 1 ms. The amplitude of the waveform covers about 1.9 divisions on the vertical axis. With the vertical axis set to 1.0 V/div, this is equal to 1.9 V. In other words, the peak-to-peak voltage is 1.9 V. The frequency of the waveform can be calculated from the period with $f = 1 / T$, giving $f = 1 / (1\times10^{-3}) = 1000$ Hz, or $f = 1$ kHz.

9.5 Function Generators

A function generator is used to generate different waveforms at various frequencies. They can be used to inject signals into circuits for testing purposes. Most function generators can generate sine wave, square wave, and triangle wave signals. Some generators can do arbitrary waveforms, which means that it is up to the user to define the output waveform. The amplitude and DC offset of the generated signal can be adjusted as required. A digital display is provided (Figure 9.9) to show the frequency and the amplitude of the generated waveform.

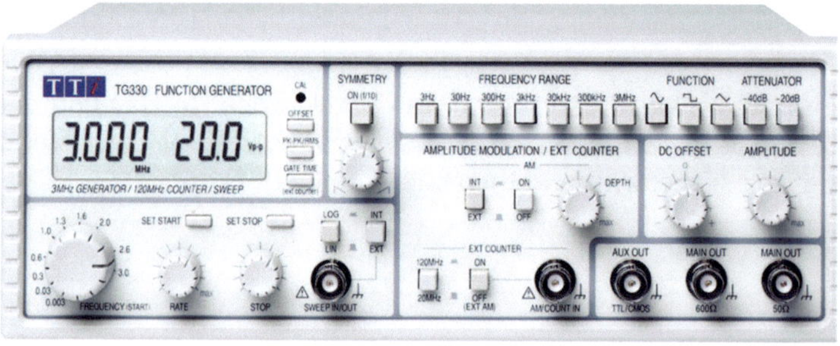

Figure 9.9 A function generator (source: www.aimtti.com).

9.6 Other Tools

In addition to the test and measuring devices, you will need additional tools in your lab. Some of these tools are described in this section.

9.6.1 Soldering Iron

A good soldering iron is necessary if you are serious about electronics. Although you can test most circuits using a breadboard, there will be occasions where you will need to solder components and build circuits. The temperature of the soldering iron used is important. If it's too hot, it can damage a component. If, on the other hand, it is too cold, then the solder will not melt well, and the solder joints will be bad. A 40-W soldering iron should be enough for most electronic projects. Of course, solder will also be required. The temperature of some soldering stations can be adjusted. It is also useful if the iron's tip can be changed, for example, for soldering smaller or larger areas. A typical, correct soldering iron is shown in Figure 9.10.

Figure 9.10 Soldering iron with adjustable temperature.

9.6.2 Desoldering Pump

Desoldering pumps are used to remove solder from printed circuit boards (PCBs) and components if, for example, you want to remove a component from a board. Figure 9.11 shows a desoldering pump.

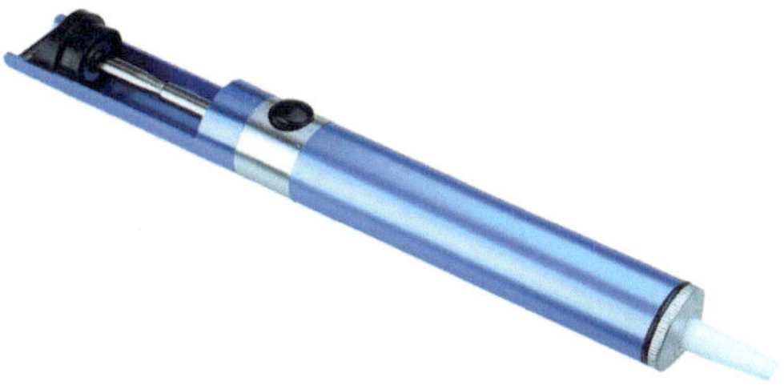

Figure 9.11 Desoldering pump.

9.6.3 Miscellaneous Tools

Small tweezers, pliers, small screwdrivers, cutters and various jumper wires will also be required in an electronic working environment.

9.6.4 Breadboard With Built-In Power Supply

You will probably be using a breadboard in your electronic projects for testing them. A battery or a DC power supply is always required in your projects. Therefore, it may be useful to purchase a breadboard with a built-in power supply. An example of such a breadboard is shown in Figure 9.12. Some devices provide fixed output voltages (e.g. +5 V), some provide adjustable output voltages.

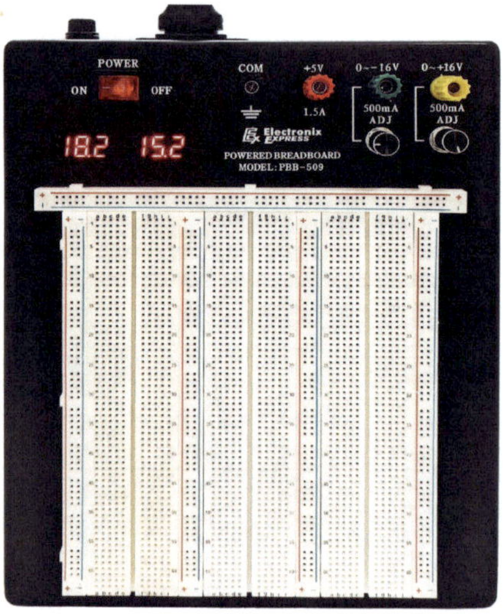

Figure 9.12 Breadboard with power supply (source: www.elexp.com).

9.6.5 Budget Tool Kits

Some manufacturers offer budget tool kits for electronic work, costing less than €50. An example of such a kit is shown in Figure 9.13. This kit contains all the tools you may need in your laboratory.

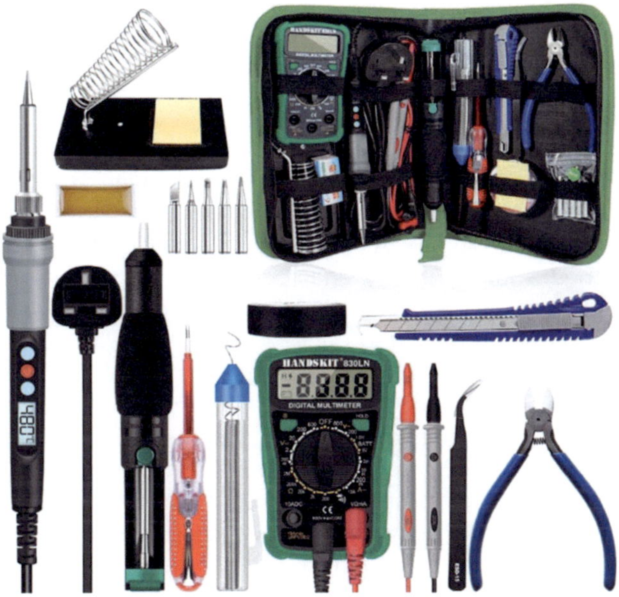

Figure 9.13 Budget tool kit (source: handskit.com).

Chapter 10 • Microcontroller Boards and MCU Programming

10.1 Overview

Early small computer systems were based on microprocessors. A microprocessor contains the ALU (Arithmetic and Logic Unit), control unit (CU), and timing circuitry. A microprocessor on its own is not useful, and it requires several external support chips such as memory, input-output, interrupt control, etc. As a result, a small computer consisted of many chips, which consumed a lot of energy and required large power supplies. Designing a microprocessor-based project was complicated, as it required many components to be connected to each other and then tested.

A microcontroller unit or MCU is a compact, integrated circuit (IC) that contains a processor core, memory, and programmable input/output peripherals. It is designed to perform specific tasks and execute embedded applications. Today, microcontrollers have largely replaced microprocessors in most systems, except for computers. The advantage of using a single chip computer is its simplicity, lower power consumption, and fewer connections. This makes hardware design and testing much easier.

10.2 Microcontroller Boards & Modules

Many microcontroller manufacturers offer development systems in the form of kits. Usually, a development board (or kit) includes a microcontroller, peripheral devices such as LEDs, buzzers, push-button switches, sockets for accessing the input-output ports, and a power supply. In general, the development boards can be connected to a PC through its USB port.

The application programs can be developed on the PC, usually using an Integrated Development Environment (IDE). An IDE includes an editor to write the program, a compiler to compile the program and generate the code the target MCU can execute, and a programmer software so that the compiled program (the executable) can be uploaded to the microcontroller's program memory.

Some IDEs also include simulators and debuggers. A simulator helps the programmer to test the program functionally before uploading it to the target microcontroller. Programmers can single step through a program or insert breakpoints and analyze the operation of the program. A debugger enables the programmer to test the program while the program is running on the target microcontroller. Again, the program can run in single steps, breakpoints can be entered at required points, and contents of program variables can be observed and changed if required.
There are many microcontroller development boards (or kits) available, manufactured and distributed by many suppliers. In this chapter, we will be briefly looking at the following popular microcontroller development boards:

- Arduino UNO R4
- ESP32-DevKitC
- Raspberry Pi Pico
- Raspberry Pi 5

The installation of the IDE will be described for each microcontroller development board. Also, a program will be given for each board to show how an LED can be connected to a microcontroller input-output port, and how the microcontroller can be programmed to flash the LED every second.

10.3 The Arduino UNO R4

Arduino is an open-source microcontroller development system, incorporating the hardware, Integrated Development Environment (IDE), and numerous software libraries. Arduino is supported by a large community of programmers, electronic engineers, enthusiasts, and academics. There are all sorts of Arduino boards. Although they are intended for different types of applications, they can all be programmed using the same IDE, and in general programs can be ported between different boards. This is one of the reasons for the popularity of the Arduino ecosystem. Arduino is also supported by many software libraries that your application can use to talk to sensors, displays and other devices. Using these libraries makes programming relatively easy and speeds up the development time. Using libraries also makes it easier to test your programs, since most libraries have already been tested and are being used by many other people.

The Arduino UNO R3 board is the most popular Arduino family member and has been with us for many years. The more recent Arduino UNO R4 is compatible with the R3, but it offers highly improved specifications compared to the R3. The UNO R4 is based on a 48-MHz 32-bit Arm Cortex-M4 processor with a lot of SRAM and program memory. Additionally, the UNO R4 has a 14-bit ADC (analog-to-digital converter) and a 12-bit DAC (digital-to-analog converter). It also supports CAN (controller-area-network), enabling the board to be connected to a CAN bus. Two versions of the board are available: the UNO R4 Minima, and the UNO R4 WiFi. In this section, we will look at the UNO R4 Minima only.

The Hardware

Figure 10.1 shows the Arduino UNO R4. At the center of the board is the RA4M1 microcontroller. At the left-hand side of the board you can see the Reset button, the USB-C connector for connecting to a PC, and the barrel jack connector for supplying external power. On the right side of the board, you can see the 6-pin ICSP connector (for In-Circuit Serial Programming and also SPI) and the 10-pin SWD/JTAG programming/debug connector. An LED ('L') is connected to port 13 (SCK). Additionally, yellow LEDs are connected to the serial TX and RX pins to indicate serial data transmission, a green LED indicates power to the board.

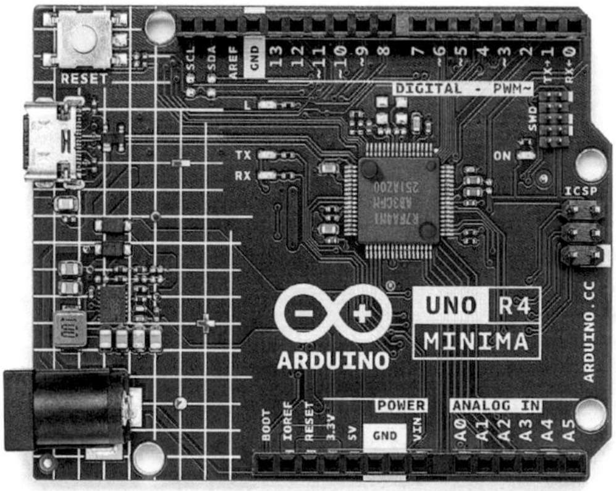

Figure 10.1 The Arduino UNO R4 Minima board (source: Arduino).

At the two long sides of the board are lined with the extension connectors. These headers have the following signals:

Short Header Power & Analog In:

A0 – A5: analog inputs
VIN: power supply connected to the barrel jack
GND
+5 V
+3.3 V
RESET: MCU reset
IOREF: reference voltage for the digital I/O pins on the extension headers.
BOOT: mode selection

Analog pins A0-A5 can also be used as digital I/O pins. Some of them have other functions:

A0: DAC output
A1: noninverting input of built-in op-amp
A2: inverting input of built-in op-amp
A3: output of built-in op-amp
A4: I²C bus SDA pin
A5: I²C bus SCL pin

Long Header Digital & PWM~:
D0 – D13: general purpose inputs and outputs (GPIO). These pins have many other functions as well:

- Interrupt inputs IRQ00 (D2) & IRQ01 (D3)
- 6× PWM outputs (D3, D5, D6, D9, D10, D11, marked on the board with '~')
- UART RX← (D0) & TX→ (D1)

- SPI MOSI (D11), MISO (D12), SCK (D13) & CS (D10)
- CAN TX (D4) & RX (D5) on the Minima; TX (D10) & RX (D13) on the WiFi

GND
AREF: reference voltage for the analog inputs (ADC)
I²C SDA (pull-up resistor)
I²C SCL (pull-up resistor)

Power is normally applied to the board through its USB-C connector when connected to a computer. Power can also be applied to the barrel connector or the VIN pin. The digital ports (GPIOs) on the microcontroller can handle currents of I_{OH} = 4 mA and I_{OL} = 8 mA. Remember that I_{OL} is the current going into the pin (sinking) when the pin is at logic LOW, and I_{OH} is the current flowing out of the pin (sourcing) when the pin is at logic HIGH. Care must be taken not to exceed the recommended current capabilities of the GPIO ports.

The IDE
Programs ('sketches') for Arduino are written in the C++ language, which is currently one of the most popular languages for programming microcontrollers. Arduino can be programmed using the Desktop Arduino IDE. This requires for the software to be installed on your PC. There is also a cloud-based IDE. The Arduino IDE consists of an editor, a compiler, a debugger, and tools to upload the compiled program to the processor on the development board. An Arduino program is called a sketch, which is compiled into machine code and is uploaded to the target processor.

The Arduino IDE has been developed over a decade and there are several versions of it. The latest stable version at the time of writing these lines is 2.3.2 which has been released in February 2024. Version 1.8.19 has been popular for many years and is still used by many programmers. Version 2 (and higher) is recommended since it is faster and easier to use than version 1.8.19. The author uses version 2 in his projects.

Installing the Arduino IDE V2
The latest version of the Arduino IDE can be installed from the following website:

https://www.arduino.cc/en/software

Select your operating system from the **DOWNLOAD OPTIONS** at the right. Click **JUST DOWNLOAD** (unless if you want to support by contributing). The author installed it on a laptop computer running Windows 10. At the time of writing this book, the latest version file name was: **Arduino-ide_2.3.2-Windows_64bit.exe**.

You now must install the Boards Package for your Arduino UNO R4 Minima. The steps are:

- Start the Arduino IDE

- Click to Open the **Boards Manager** at the top left of the screen (Figure 10.2)

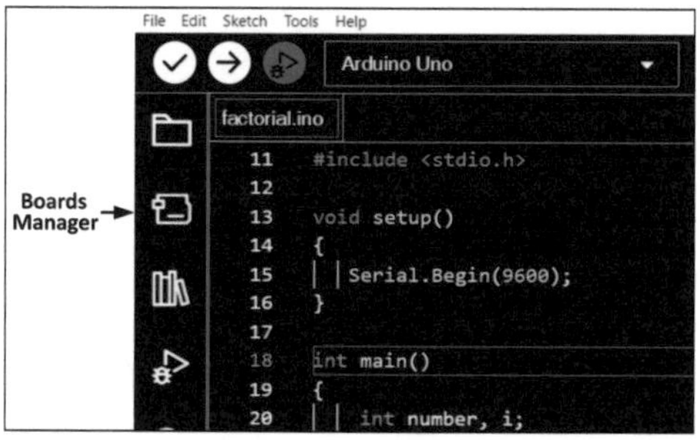

Figure 10.2 Open the Boards manager.

- Search for Arduino UNO R4 Boards (Figure 10.3) and click INSTALL to install it. At the time of writing this book, the version was 1.0.1.

Figure 10.3 Install the Arduino UNO R4 boards package.

- Click **Boards Manager** to close the left window.

- You should be able to select the Arduino UNO R4 board from the board selector at the top left of the screen (Figure 10.4). Connect your Arduino UNO R4 Minima to your PC via a USB-C cable.

- Click **Tools → Port** and select the serial port connected to your Arduino UNO R4.

Figure 10.4 Select Arduino UNO R4 Minima.

You are now ready to develop programs and upload them to your Arduino UNO R4 Minima development board.

Before looking at some example programs, it is worthwhile to learn some of the *commonly used* menu options offered by the IDE.

File: Open existing programs, save programs, open example programs, and set the IDE working environment (e.g. Preferences...)

Edit: Select, cut & paste text, go to a specified line, change indentation and font size, and search for text in a file.

Sketch: Compile your program, upload the compiled program to the target processor, include libraries, add files and some other options that you will not be using right now.

Tools: Manage libraries, open the serial monitor and serial plotter, select and configure the development board that you will be using, and burn a new bootloader.

Help: Display various help items on selected topics. Additionally, it displays the version number of the currently used IDE.

10.3.1 Project 1 – Blink an External LED

Description: In this project, you will be connecting an LED to the development board and blink it every second. The aim of this project is to show how an LED can be connected to the development board and how a program can be uploaded to the development board.

Block diagram: Figure 10.5 shows a high-level overview of the project.

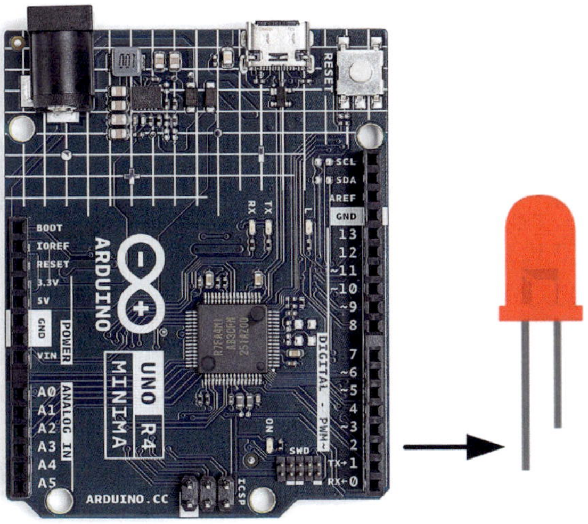

Figure 10.5 High-level overview of the project

Circuit diagram: LEDs can be connected to the development board either in current-sourcing mode or current-sinking mode. In current-sourcing mode (Figure 10.6), one pin of the LED is connected to the port through a current limiting resistor and the other pin is connected to the supply ground. In this mode, the LED is ON when a logic 1 (HIGH, TRUE) is applied to it. In current-sinking mode (Figure 10.7), one pin of the LED is connected to the positive power supply and the other pin to the port through a current limiting resistor. In this mode, the LED is ON when a logic 0 (LOW, FALSE) is applied to it.

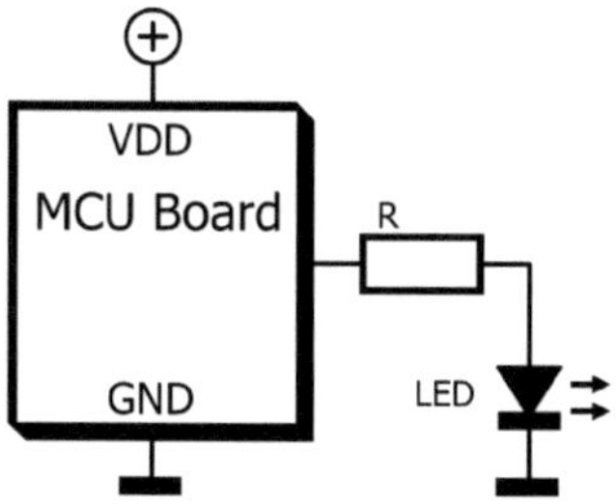

Figure 10.6 LED connected in current-sourcing mode.

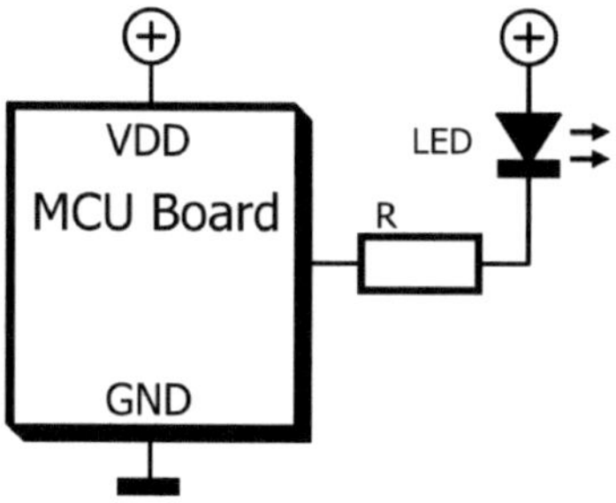

Figure 10.7 LED connected in current-sinking mode.

The value of the current limiting resistor can be calculated as follows. Assuming the desired current through the LED is 3 mA and the voltage drop across the LED for this current is 1.8 V (typical value for a red LED; you can find this value in the LED's datasheet), then:

R = (5 – 1.8) / 0.003 = 1067 Ω. We can use a standard-value resistor of 1 kΩ.

In this project, the LED is connected in current-sourcing mode as shown in Figure 10.8.

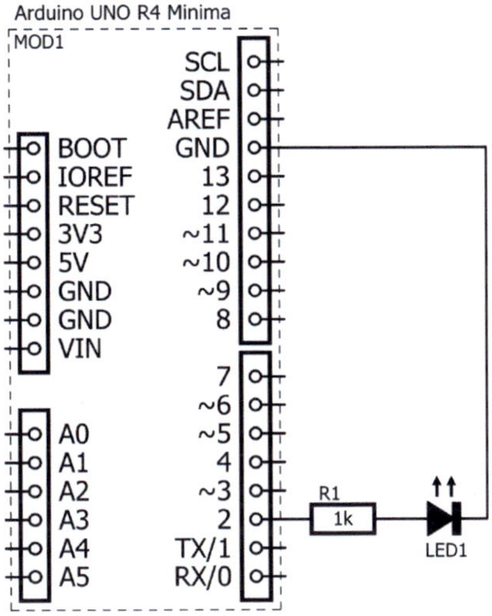

Figure 10.8 Connecting the LED in current-sourcing mode

Construction: The circuit is constructed on a breadboard and connections are made using jumper wires. Figure 10.9 shows the wiring diagram of the circuit.

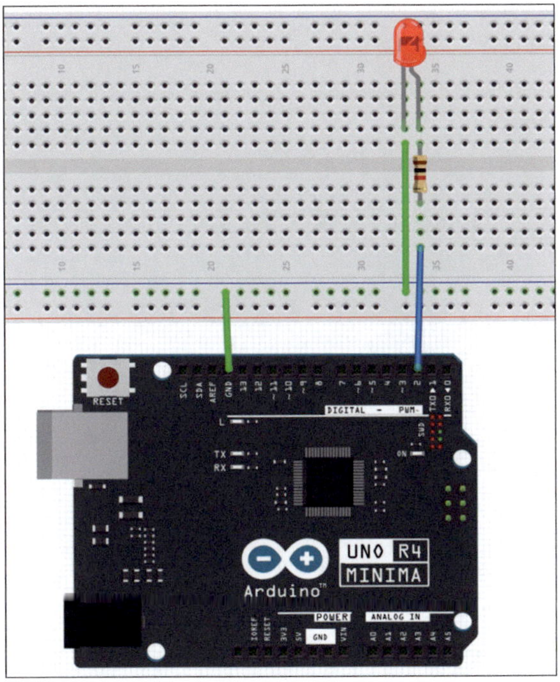

Figure 10.9 Breadboard wiring diagram of the circuit.

Program listing: Figure 10.10 shows the program listing (Program: **LEDext**). At the beginning of the program, ON and OFF are defined, and the LED is assigned to I/O port 2.

The setup() function is run once at the beginning of the program. Here, the LED pin is configured as an output. The loop() function repeats itself continuously. Here, the LED is turned ON, then the program waits one second before turning the LED OFF. Then the program waits again one second before starting over. The result is that the LED toggles its state every second (therefore, it blinks at a rate of 0.5 Hz).

The steps to write and upload the program are given below:

- Start the IDE
- Write the program as in Figure 10.10.
- Click **Sketch → Verify/Compile** to compile your program. Make sure there are no error messages. If there are error messages, you must go back to the program and correct them.
- Connect the Arduino board to your computer. Click **Tools → Board** and select your board. Click **Tools → Port** and select the serial port assigned to your board.
- Click **Sketch → Upload** to upload the compiled code to the program memory of the board.
- Press the reset button on your hardware. You should see the LED change state every second.

```
//----------------------------------------------------------------------
//                        BLINKING AN EXTERNAL LED
//                        ========================
//
// This program blinks the LED connected to port 2 of the board
//
// Author: Dogan Ibrahim
// File  : LEDext
// Date  : December, 2023
//----------------------------------------------------------------------
#define ON HIGH                                    // Define ON
#define OFF LOW                                    // Define OFF
int LED = 2;                                       // LED at port 2

void setup()
{
  pinMode(LED, OUTPUT);                            // Configure LED as output
}

void loop()
```

```
{
  digitalWrite(LED, ON);                          // LED ON
  delay(1000);                                    // 1 second delay
  digitalWrite(LED, OFF);                         // LED OFF
  delay(1000);                                    // 1 second delay
}
```

Figure 10.10 Program: LEDext

10.4 The ESP32-DevKitC

The ESP32 is a low-cost, low-power microcontroller family with built-in Wi-Fi and Bluetooth. In its original version, the ESP32 has two 32-bit Xtensa LX6 processors operating at up to 240 MHz. It has a good amount of SRAM and features rich peripheral support such as ADC, DAC, touch sensors, and popular communication interfaces such as SPI, I²S, I²C, UART, CAN bus, etc. Since its first release, several different models of the original ESP32 processor have been introduced.

The basic specifications of the ESP32 MCU are summarized below:

- Dual-core Tensilica Xtensa LX6 RISC processor
- Operation speed 160 MHz to 240 MHz
- 520 KB SRAM
- 448 KB Flash memory
- 16 KB SRAM (in RTC)
- IEEE 802.11 b/g/ne/I Wi-Fi
- Bluetooth 4.2
- 18-channel 12-bit ADC
- 2-channel 8-bit DAC
- 10 touch sensors
- 36× GPIO
- 4× SPI
- 2× I²C
- 2× I²S
- 3× UART
- CAN bus 2.0
- SD memory card support
- 2.2 V – 3.36 V operation
- RTC timer and watchdog
- 16-channel PWM
- Ethernet interface
- Internal 8 MHz and RC oscillator
- External 2 MHz–60 MHz and 32 kHz oscillator
- Cryptographic hardware acceleration (AES, HASH, RSA, ECC, RNG)
- IEEE 802.11 security features
- 5 μA sleep current

The ESP32-DevKitC is one of the most popular ESP32 modules around, and we will use it in this chapter.

ESP32-DevKitC Hardware

ESP32-DevKitC is a small ESP32-based board developed and manufactured by Espressif. As shown in Figure 10.11, the board measures 54.4 mm by 27.9 mm, and it fits on a breadboard.

Figure 10.11 The ESP32-DevKitC module.

The board has two connectors located along each side of the board for GPIO, clock, and power line interfaces. Each connector has 19 pins. The two connectors carry the following signals:

Left Connector	**Right Connector**
3.3 V	GND
EN	O23
SVP	IO22
SVN	TXD0
IO34	RXD0
IO35	IO21
IO32	GND
IO33	IO19
IO25	IO18
IO26	IO5
IO27	IO17
IO14	IO16
IO12	IO4
GND	IO0
IO13	IO2
SD2	IO15
SD3	SD1
CMD	SD0
5 V	CLK

The board has a micro-USB connector for connecting it to a computer. The board also receives its power from the USB port. Standard 5 V from the USB port is converted into

3.3 V on the board. In addition, two buttons are provided on the board named EN and BOOT, having the following functions:

EN: Pressing this button resets the board.

BOOT: Boot mode selection. The board is in normal operation mode when the button is not pressed. Pressing and holding down this button while at the same time pressing the EN button starts programming mode, which allows uploading of new firmware to the processor through the USB serial port.

The board operates from a power supply of +3.3 V (the absolute maximum is +3.6 V). It is recommended that the current capacity of each pin should not exceed 6 mA (the absolute maximum current capacity is 12 mA). It is therefore important to use proper current limiting resistors while driving external loads such as LEDs.

Installing the Arduino IDE for the ESP32-DevKitC

The Arduino IDE we have introduced in the previous section can also be used to program the ESP32 and to upload the compiled code.

The steps to install the ESP32 on Arduino IDE are as follows:

- Download and install the latest version of Arduino IDE from the following website:

  https://www.arduino.cc/en/Main/Software

Open your Arduino IDE and click **File → Preferences** to open the Preferences window. Locate text box **Additional Board Manager URLs** at the bottom of the window and enter the following line. If the text box contains another URL, add the new URL after it, separating them with a comma:

  https://dl.espressif.com/dl/package_esp32_index.json

- Click **OK**
- Click **Tools → Boards → Board Managers** window and search for ESP32.
- Click the **Install** button. You should see the Installed message. **Close** the window.
- We can now write programs for the ESP32 processor and then upload them to the board.

10.4.1 Project 2 – Blinking an LED With the ESP32

Description: In this project, an LED is connected to GPIO port 23 of the ESP32-DevKitC. The LED changes state every second. The aim of this project is to show how an LED can be connected to a GPIO port and how it can be turned on and off. The project also describes how the program can be uploaded to the ESP32 program memory using the Arduino IDE.

Block diagram: Figure 10.12 shows the high-level overview of the project.

Figure 10.12 High-level overview of the project.

Circuit Diagram: Figure 10.13 shows the pin layout of the ESP32-DevKitC. The circuit diagram of the project is shown in Figure 10.14. The LED is connected to GPIO port 23 (second pin at the top right-hand side) through a 330 Ω current limiting resistor.

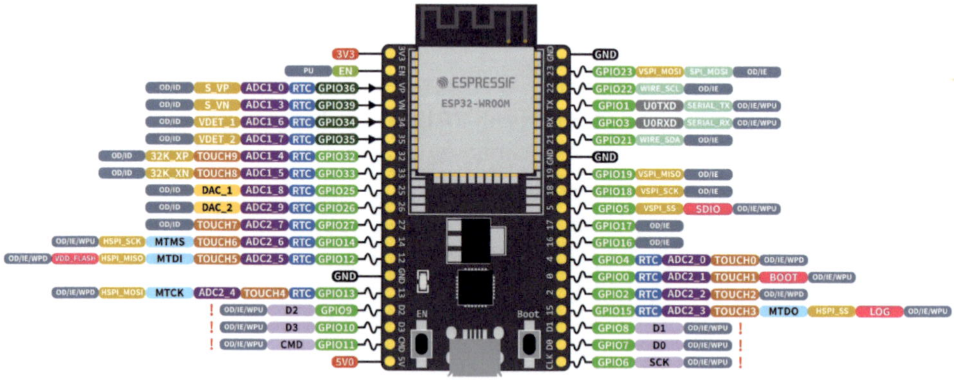

Figure 10.13 Pin layout of the ESP32-DevKitC (source: Espressif)

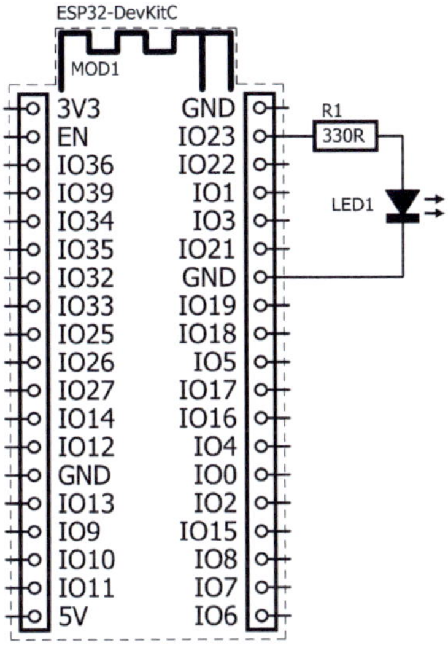

Figure 10.14 Circuit diagram of the project.

Construction: The ESP32-DevKitC board is mounted on a breadboard as shown in Figure 10.15. The LED is connected to the board through a current limiting resistor.

Figure 10.15 The project on a breadboard.

Program Listing:
The program listing of the project is very simple and is shown in Figure 10.16 (program: **FlashLED**).

```
/*****************************************************************
 *                        FLASHING LED
 *                        ============
 *
 * In this program an LED is connected to port GPIO23 of the
 * ESP32 DevKitC. The program flashes the LED every second
 *
 * Program: FlashLED
 * Date    : December, 2023
 *****************************************************************/
#define LED 23

void setup()
{
    pinMode(LED, OUTPUT);
}


void loop()
{
    digitalWrite(LED, HIGH);
    delay(1000);
    digitalWrite(LED, LOW);
    delay(1000);
}
```

Figure 10.16 Program listing for blinking an LED with the ESP32.

Program Description: At the beginning of the program, the variable name LED is assigned to GPIO port 23. Then the LED port is configured as an output. The main program is run continuously in a loop, where inside this loop the LED is turned ON (HIGH) and OFF (LOW) with a one-second delay between each change of state.

The steps to compile and upload the program are as follows:

- Start the IDE.
- Write the program as in Figure 10.16.
- Select **Tools → Board → ESP32 Dev Modul**e in the IDE.
- Click **Sketch → Verify/Compile** to compile your program. Make sure there are no error messages. If there are, go back to the editor and correct them.
- Connect the ESP32-DevKitC hardware to your PC, click **Tools → Port** and select the serial port assigned to your board.

- Click **Sketch → Upload** to upload the compiled code to the program memory of the ESP32 module. Note that you might have to press and hold the BOOT button on the ESP32-DevKitC during the program upload process.

- Press the reset button on the board. You should see the LED change state every second.

10.5 The Raspberry Pi Pico

The Raspberry Pi Pico (we'll call it just Pico from now on) is a microcontroller module developed by Raspberry Pi. The Pico is a low-cost module based on the RP2040 dual-core Arm Cortex-M0+ MCU. Figure 10.17 shows the top view of the Pico. In the middle of the board sits the tiny 7 mm by 7 mm RP2040 housed in a QFN-56 package. A strip of 20 holes (0.1″ pitch) running along each long side of the board allows the MCU to connect to the outside world. You can solder pins or wires to these holes. If you solder pin headers, then the board can be plugged on a breadboard (note that the Pico can also be bought with pin headers mounted). Next to these holes, you will see contacts that look like holes cut in half. These are so-called castellated contacts, and they are intended for soldering the board directly onto another board. This saves headers and keeps the assembly flat.

Figure 10.17 Top view of the Raspberry Pi Pico (source: Raspberry Pi).

On the right short edge of the board, there is the micro-USB-B connector for providing power to the board and for programming the MCU. Next to the USB port is an LED that can be used during program development. The button named BOOTSEL is used during programming of the microcontroller. At the other (left) short edge of the board are three contacts to which you can attach a Serial Wire Debug (SWD) debugger.

Figure 10.18 shows the rear of the Pico hardware module. Here, all the GPIO pins are identified with letters and numbers.

Figure 10.18 Bottom side of the Pico (source: Raspberry Pi).

The specifications of the Pico are as follows:

- RP2040 dual-core 32-bit Arm Cortex-M0+ operating at 133 MHz
- 2 MB Q-SPI Flash memory

- 264 KB SRAM
- 26× GPIO (+3.3 V compatible)
- 3× 12-bit ADC pins
- 2× UART
- 2× I^2C
- 2× SPI
- 16 PWM channels
- 1× Timer (with 4 alarms)
- 1× Real-Time Counter
- Temperature sensor
- LED (on port GP25)
- Serial Wire Debug (SWD) port
- Micro-USB port (USB 1.1) for power (+5V) and data (programming)
- Suitable for programming in MicroPython & C/C++
- Drag & drop programming over USB

The Pico's GPIO ports are +3.3 V compatible, and it is therefore important to be careful not to exceed this voltage when interfacing external devices to these pins. As they are not 5 V-tolerant, a 5-V-to-3.3-V level shifter is required to interface to 5 V signals. This can be as simple as a resistive voltage divider, or you can use a special IC for it.

The Pico can be powered in several ways. The simplest method is to plug the micro-USB port into a 5 V power source, such as the USB port of a computer or a 5 V USB power adapter. Another method is by connecting a (1.8 V to 5.5 V) power source to the VSYS pin (e.g. using batteries or a power supply).

10.5.1 Pinouts of the RP2040 and the Pico

Figure 10.19 shows the RP2040 MCU pinout, which is housed in a 56-pin package. The Pico module pinout is shown in Figure 10.20 in detail. As you can see from the figure, most pins have multiple functions. For example, GPIO0 (pin 1) can also be the UART0 TX pin, the I^2C0 SDA pin, or the SPI0 RX pin.

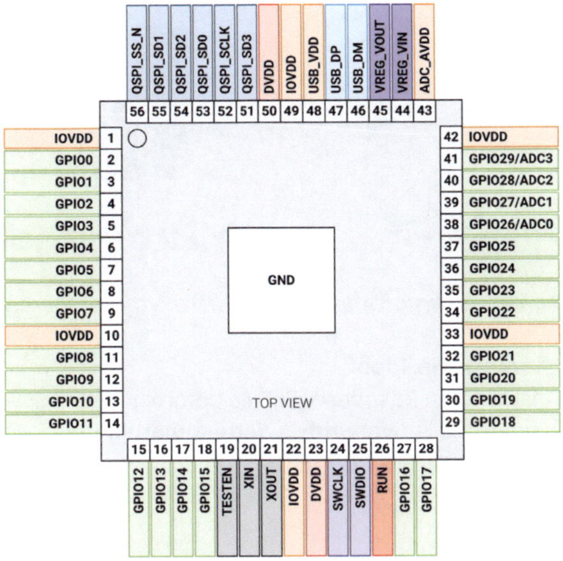

Figure 10.19 RP2040 pinout (source: Raspberry Pi).

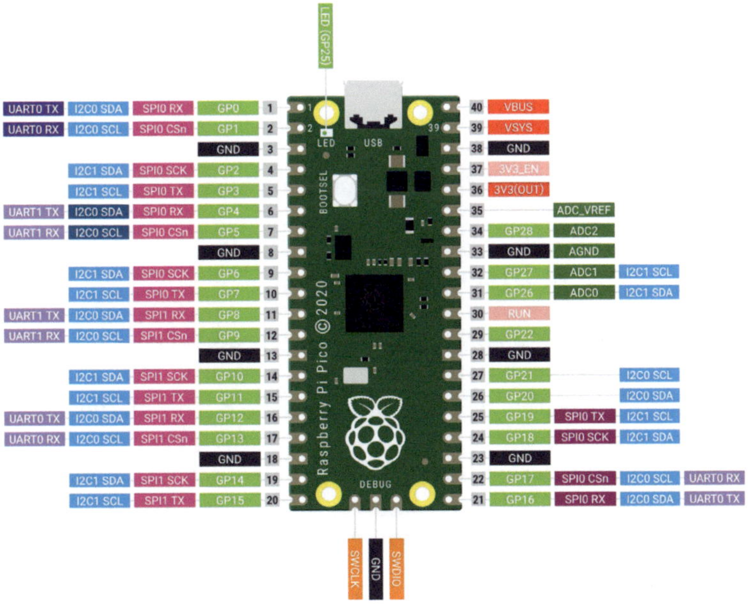

Figure 10.20 Pico pinout (source: Raspberry Pi).

10.5.2 Pico Programming

The Raspberry Pi Pico is usually programmed in C/C++ or MicroPython. Although the Pico is by default set up for use with C/C++, many beginners find it easier to use MicroPython, which is a version of the Python programming language adapted for microcontrollers.

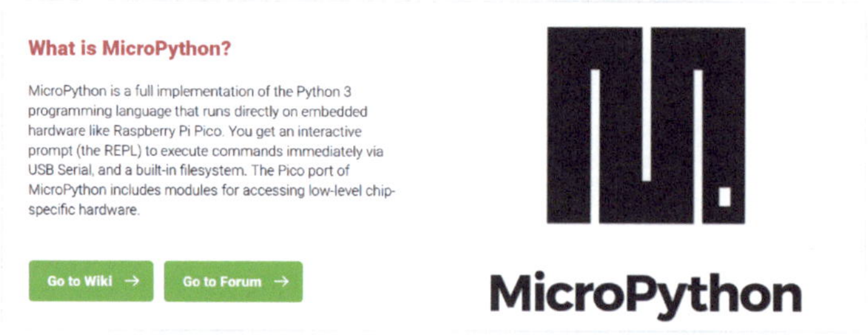

Figure 10.21 The Pico can be programmed with MicroPython (source: micropython.org).

Installing MicroPython on the Pico

MicroPython must be installed on Raspberry Pi Pico before it can be used. Once it is installed, it stays on your Pico, unless it is overwritten with something else. Installing MicroPython requires downloading a file from the Internet.

The steps are as follows:

- Download the Raspberry Pi Pico MicroPython UF2 file to a folder (e.g. Downloads) on your computer from this link:
  https://www.raspberrypi.com/documentation/microcontrollers/micropython.html

(At the time of writing this book the file was called **rp2-pico-latest.uf2**).

- Push and hold down the **BOOTSEL** button on your Pico.
- Connect your Pico to the USB port of your PC using a micro-USB cable while holding down the **BOOTSEL** button.
- Wait a few seconds before releasing the **BOOTSEL** button.
- You should see the Pico appear as a removable drive with the name **RPI-RP2**
- Drag and drop the downloaded MicroPython UF2 file onto the RPI-RP2 drive. Your Pico will reboot. When ready, it is running MicroPython. Powering down the Pico will not remove MicroPython from its memory.

Using the Thonny IDE From the PC

Programs for the Pico can be written and the uploaded to its memory using the Thonny IDE. In this section, you will be learning how to use the Thonny IDE on your computer.

First, you must install Thonny on the computer. The steps are:

- Go to the Thonny.org website at
  https://thonny.org/

- Download and install Thonny for your operating system (see Figure 10.22)

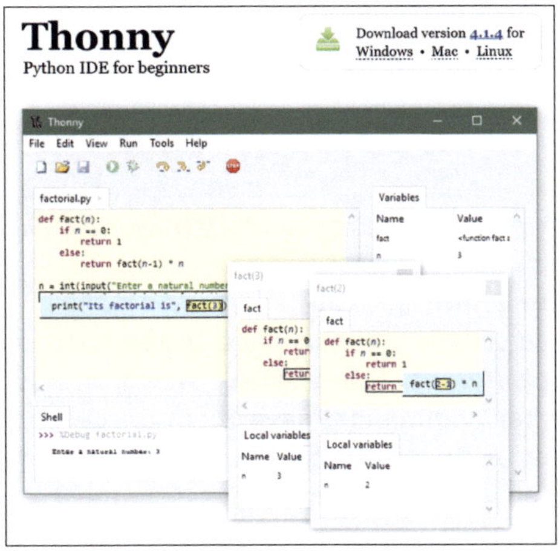

Figure 10.22 Download Thonny for your operating system (source: thonny.org).

- Double-click the Thonny icon to start the program.

- The startup screen of Thonny on your PC is shown in Figure 10.23.

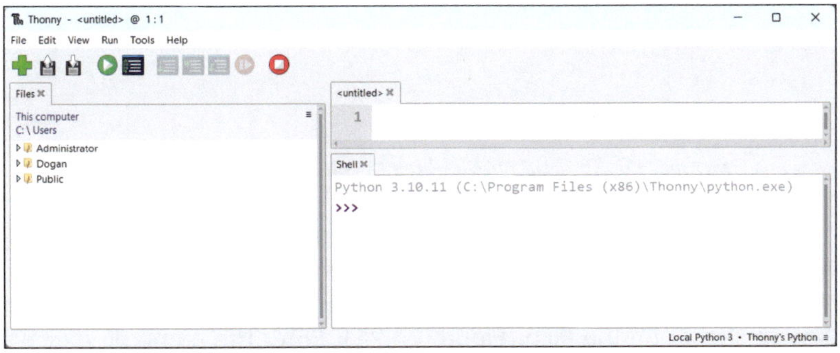

Figure 10.23 The Thonny IDE.

- Click on the label **Local Python** 3 at the bottom right-hand corner of the window (the text may be different on your computer) and click to select **MicroPython (Raspberry Pi Pico)**

You are now ready to write your MicroPython programs.

Let us write a simple Python command to test the installation of MicroPython on our Pico. Enter the following command after the '>>>' prompt in the **Shell** window:

```
print("hello from Thonny on PC")
```

You should see the message displayed as shown in Figure 10.24.

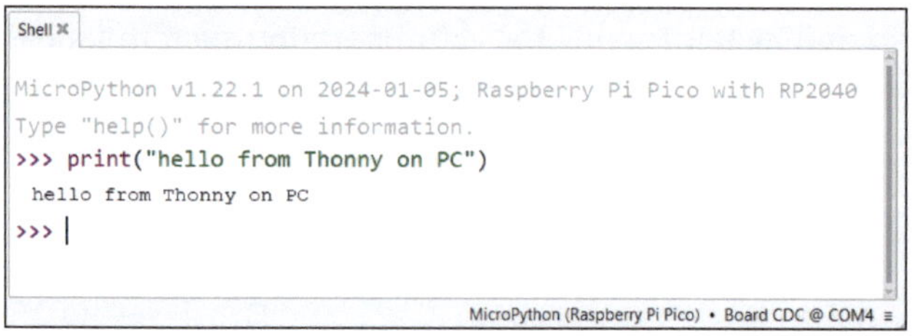

Figure 10.24 MicroPython communication with the Pico established.

Writing a Program Using Thonny

In the previous section, you saw how to execute a statement in real time using the Thonny interactive shell. This shell is a so-called **REPL**, which stands for "Read-Evaluate-Print Loop." It allows you to enter commands, have them executed, and see the results in as you go. This is practical for testing small code snippets, experimenting with programming constructs, and learning programming languages. For most applications, however, you cannot use a REPL shell, and you will have to write a program instead. Therefore, as an example, let's write and execute a simple single-line program to display a message:

- Enter the single-line program in the editor window of Thonny, as shown in Figure 10.25.

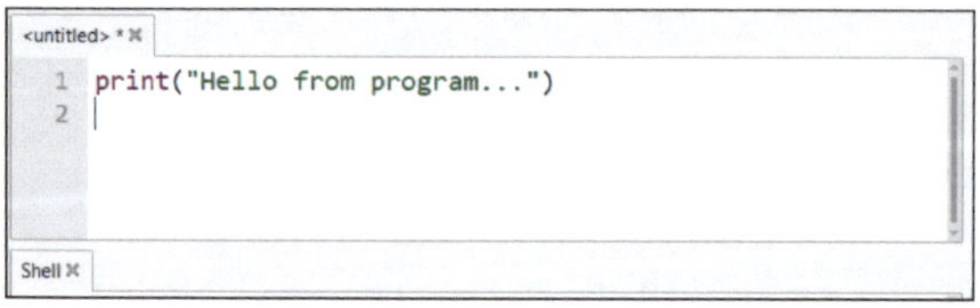

Figure 10.25 Our single-line program test program.

- Click **File** followed by **Save As** and give a name to your program. e.g. **FirstProg.py**. You have the option of storing the program either on your PC (**This computer**) or on the Pico (**Raspberry Pi Pico**). Click **Raspberry Pi Pico** to save it on the Pico (Figure 10.26). Enter the name of your program (**FirstProg.py**) and click **OK** (notice that the file is saved with the extension **.py**).

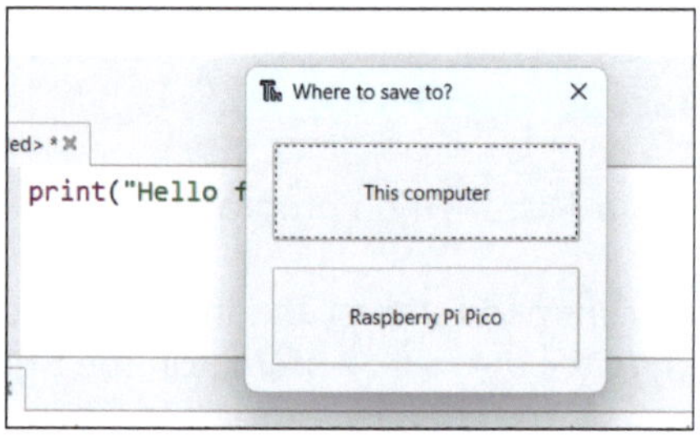

Figure 10.26 Click the Raspberry Pi Pico button to save your program to the Pico.

- Click the green arrow icon in the toolbar at the top of the Thonny window to run your program. The output of the program will be displayed in the Shell window, as shown in Figure 10.27.

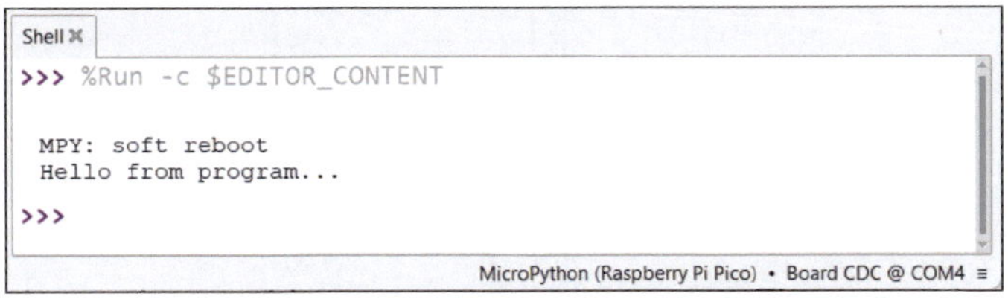

Figure 10.27 Output of the single-line program.

10.5.3 Project 3 – Blinking an LED With MicroPython

Description: In this project, an LED is connected to the Pico. As before, the LED will change state every second (i.e. blink at 0.5 Hz). The aim of this project is to show how to connect an LED to the Pico and how to program the Pico in MicroPython to control the LED.

Block diagram: Figure 10.28 shows the high-level overview of the project.

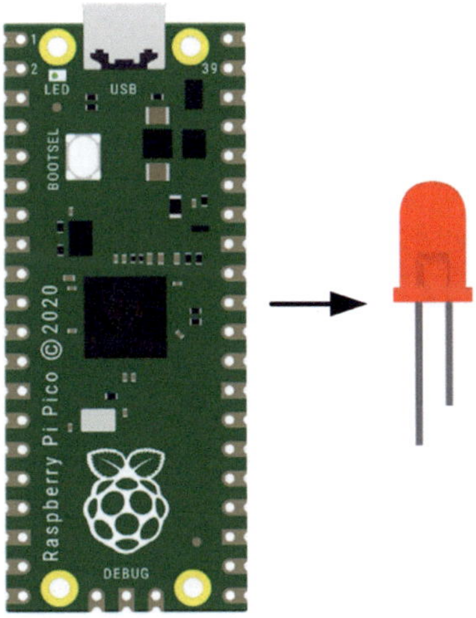

Figure 10.28 High-level overview of the project

Circuit diagram: The circuit diagram of the project is shown in Figure 10.29 where the LED is connected to port pin GP1 of the Pico through a 470 Ω current-limiting resistor.

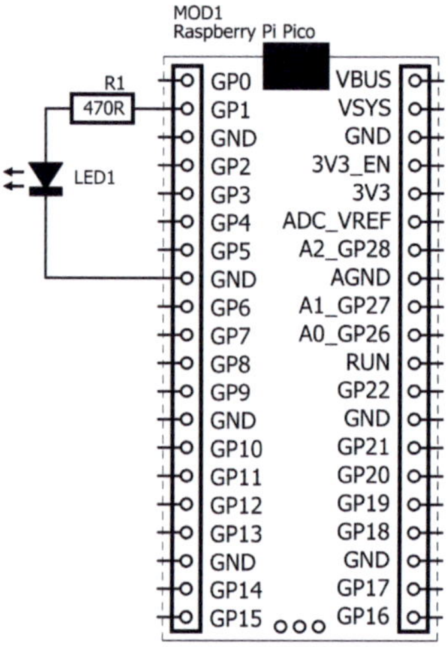

Figure 10.29 Circuit diagram of the project.

Program listing: Figure 10.30 shows the program listing (Program: **ExtFlash.py**). At the beginning of the program, the LED at GP1 is configured as an output. An infinite **while** loop then runs until the Stop button is clicked in Thonny. Inside this loop, the LED is turned ON and OFF with a delay of one second between each state change.

```
#-----------------------------------------------------------
#                          FLASHING AN EXTERNAL LED
#                          ========================
#
# In this program an external LED is connected to port pin
# GP1 (pin 2). The LED is flashed every second
#
# Author: Dogan Ibrahim
# File  : ExtFlash.py
# Date  : December, 2023
#------------------------------------------------------------
from machine import Pin
import utime

LED = Pin(1, Pin.OUT)               # LED at GP1

while True:                         # DO FOREVER
    LED.value(1)                    # LED ON
    utime.sleep(1)                  # Wait 1 second
    LED.value(0)                    # LED OFF
    utime.sleep(1)                  # Wait 1 second
```

Figure 10.30 Program: ExtFlash.py

Construction: Figure 10.31 shows how to build the project on a breadboard. GP1 is the second pin. It is a bit confusing that it is labeled '2' on the top side of the Pico.

Figure 10.31 The project can be built easily on a breadboard.

The **machine** module supports the following functions (we will see in later chapters how to use these functions):

- Pin
- Timer
- ADC
- I²C and Soft I²C
- SPI and SoftSPI
- WDT
- PWM
- UART

The general format of the machine.Pin function is (for more information, see https://docs.micropython.org/en/latest/library/machine.Pin.html):

```
machine.Pin(pin, mode, pull, value, alt)
```

The parameters of a pin can be reinitialized using the following function:

```
Pin.init(pin, mode, value, drive, at)
```

Where **pin** is the pin's number.

The parameter **mode** can take the following values:

Pin.IN	pin is configured as an input.
Pin.OUT	pin is configured as an output.
Pin.OPEN_DRAIN	pin is configured as an open-drain output.
Pin.ALT	pin is configured as an alternative function.

The parameter **pull** can take the following values:

NONE	No internal pull-up or pull-down resistors.
Pin.PULL_UP	Internal pull-up resistor enabled.
Pin.PULL_DOWN	Internal pull-down resistor enabled.

The parameter **value** is valid only for Pin.OUT and Pin.OPEN_DRAIN modes, and it specifies the initial value of the output pin (if specified)

The parameter **alt** specifies an alternate function for the pin (port dependent). This parameter is valid only for PIN.ALT and Pin.ALT_OPEN_DRAIN modes.

A pin can be set/reset using one of the following functions:

Pin.value(1)	Set pin to logic 1.
Pin.value(0)	Set pin to logic 0.

Some other useful **machine** functions are:

machine.reset()	Reset the device (same as pressing a reset button).
machine.reset_cause)	Obtain the cause of the reset.
machine.disable_irq()	Disable interrupt requests.
machine.enable_irq()	Enable interrupt requests.
machine.freq()	Obtain the CPU frequency.

10.6 The Raspberry Pi

The Raspberry Pi is a series of small, single-board computers developed by Raspberry Pi. These computer boards have gained popularity for their versatility, low cost, and community support. The Raspberry Pi 5 is the most recent and powerful model at the time of writing these lines. Key features of the Raspberry Pi include:

- It is a complete computer with the size of a credit card.
- GPIO (General-Purpose Input/Output): it has a 2×20-pin extension connector that allows connecting external electronic components and peripherals, making it suitable for hardware projects and prototyping.
- It runs the 64-bit Raspberry Pi OS, an operating system based on Debian Linux.

The Raspberry Pi 5 is based on a 64-bit quad-core Cortex-A76 Arm processor running at 2.4 GHz, shouldered by an 800 MHz GPU (Graphics Processor Unit). Compared to earlier versions of the board, the Raspberry Pi 5 has a number of interesting new features such as a PCIe connector, a power button, a built-in real-time clock (RTC), a dual camera interface, and more.

Figure 10.32 shows the front view of the Raspberry Pi 5 with the components labelled for reference.

Figure 10.32 The Raspberry Pi 5 (source: Raspberry Pi).

The Raspberry Pi 5 gets rather warm, and it is recommended to use a cooler to lower the CPU temperature. Figure 10.33 shows the Raspberry Pi 5 with the active cooler installed. The fan of the cooler operates automatically when the CPU temperature goes above 50 °C.

Figure 10.33 Raspberry Pi 5 with active cooler (source: Raspberry Pi).

10.6.1 Installing the Raspberry Pi Operating System

The Raspberry Pi operating system is available either on a pre-installed microSD card, or you can download it and install it yourself on a blank microSD card. In this section, we will be using a pre-installed SD card.

Using a Pre-Installed SD Card

The pre-installed Raspberry Pi operating system is available in various microSD cards sizes. In this section, we will use a 32 GB microSD card. Additionally, an HDMI-compatible monitor, a keyboard, and a mouse are needed. The author's hardware setup is shown in Figure 10.34.

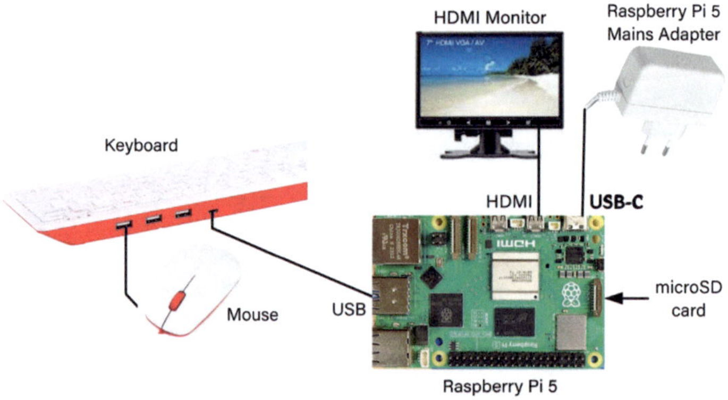

Figure 10.34 The author's Raspberry Pi 5 setup. The mouse connects to the keyboard, but it can also be connected to the Raspberry Pi directly.

The steps are as follows:

- Insert the pre-installed microSD card into your Raspberry Pi.
- Connect all the devices as in Figure 10.34.
- Connect the Raspberry Pi power adapter to the mains.
- You should see the Raspberry Pi booting. The first time it will ask you various questions to set up the device, such as the username, password, Wi-Fi network to connect to with the required credentials, etc. In this section, the username is set to **pi** and the password is **raspberry**.
- The Raspberry Pi will boot in Desktop mode and will display the default screen. You can press Ctrl+Alt+F1 at any time to change to the Console mode.

Accessing the Raspberry Pi Console Remotely

In many applications, you may want to access your Raspberry Pi from your computer. This requires enabling SSH on the Raspberry Pi and then using a terminal emulation program on your computer. The steps to enable SSH are as follows:

- Make sure you are in Console mode
- Type: **sudo raspi-config**

- Move down to **Interface Options**
- Highlight **SSH** and press Enter (Figure 10.35)

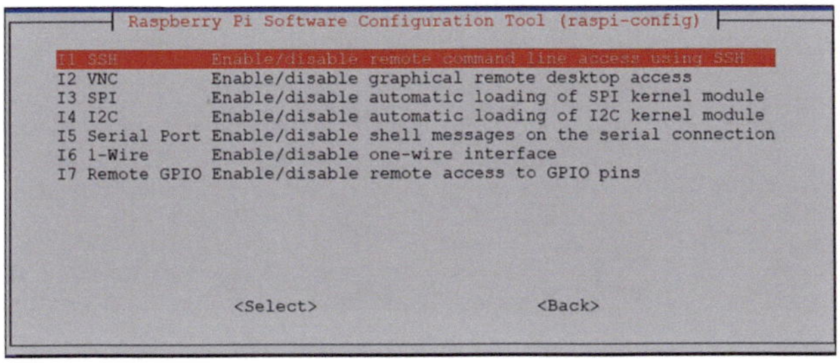

Figure 10.35 Highlight SSH and press Enter to activate it.

- Click **Yes** to enable SSH
- Click **OK**
- Move down and click **Finish**

You will now have to install a terminal emulation program on your computer. The one used by the author is the popular PuTTY. Download PuTTY from https://putty.org

- PuTTY does not require installation. Simply double-click the file to run it. You should see the PuTTY startup screen as in Figure 10.36.

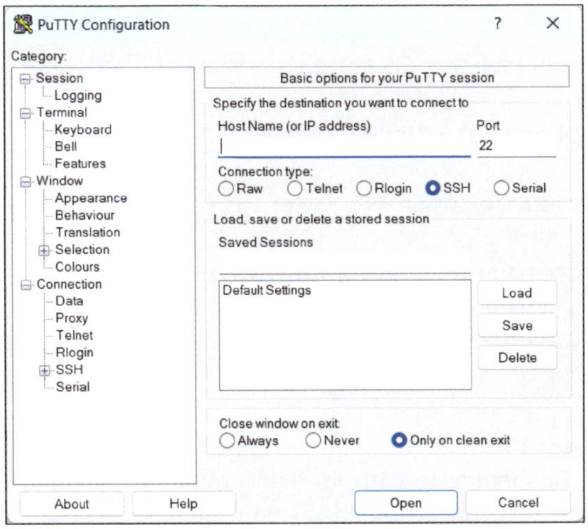

Figure 10.36 PuTTY startup screen.

- Make sure that the **Connection type** is SSH and enter the IP address of your Raspberry Pi. You can obtain the IP address by entering the command **ifconfig** in the console on the Raspberry Pi (Figure 10.37).

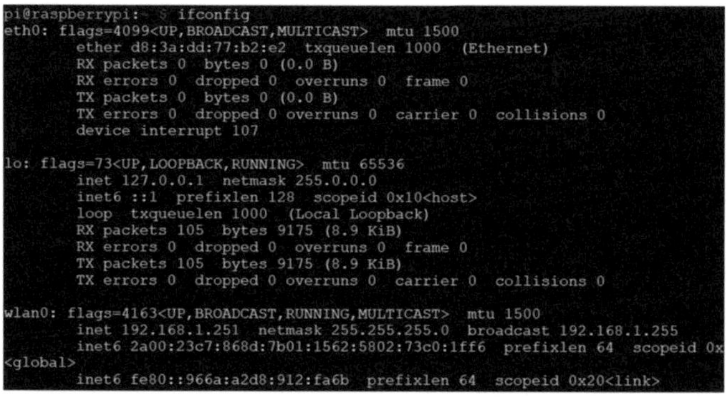

```
pi@raspberrypi:~ $ ifconfig
eth0: flags=4099<UP,BROADCAST,MULTICAST>  mtu 1500
        ether d8:3a:dd:77:b2:e2  txqueuelen 1000  (Ethernet)
        RX packets 0  bytes 0 (0.0 B)
        RX errors 0  dropped 0  overruns 0  frame 0
        TX packets 0  bytes 0 (0.0 B)
        TX errors 0  dropped 0 overruns 0  carrier 0  collisions 0
        device interrupt 107

lo: flags=73<UP,LOOPBACK,RUNNING>  mtu 65536
        inet 127.0.0.1  netmask 255.0.0.0
        inet6 ::1  prefixlen 128  scopeid 0x10<host>
        loop  txqueuelen 1000  (Local Loopback)
        RX packets 105  bytes 9175 (8.9 KiB)
        RX errors 0  dropped 0  overruns 0  frame 0
        TX packets 105  bytes 9175 (8.9 KiB)
        TX errors 0  dropped 0 overruns 0  carrier 0  collisions 0

wlan0: flags=4163<UP,BROADCAST,RUNNING,MULTICAST>  mtu 1500
        inet 192.168.1.251  netmask 255.255.255.0  broadcast 192.168.1.255
        inet6 2a00:23c7:868d:7b01:1562:5802:73c0:1ff6  prefixlen 64  scopeid 0x
<global>
        inet6 fe80::966a:a2d8:912:fa6b  prefixlen 64  scopeid 0x20<link>
```

Figure 10.37 Use the ifconfig command to obtain the Raspberry Pi's IP address. Here the IP address is 192.168.1.251 (see under wlan0)

- Click **Open**

- The first time you connect with PuTTY to your Raspberry Pi, you may get a security message. Click **Yes** to accept this security alert.

- You will then be prompted to enter the username and password.

You can now enter all console commands from your computer.

- To change your password, enter the following command:

  pi@raspberrypi: ~ $ **passwd**

- To restart the Raspberry Pi, enter the following command:

  pi@raspberrypi: ~ $ **sudo reboot**

- To shut down the Raspberry Pi, enter the following command:

  pi@raspberrypi: ~ $ **sudo shutdown –h now**

Never shutdown the Raspberry Pi by pulling its power cable, as this may result in the corruption or loss of files.

Accessing the Desktop Remotely

If you are using your Raspberry Pi with its own keyboard, mouse, and display, then you can skip this section. If, on the other hand, you want to access your desktop remotely from another computer over the network, you will find that SSH services cannot be used. The easiest and simplest way to access your desktop remotely is by using the VNC (Virtual Network Connection) client and server programs. The VNC server runs on the Raspberry Pi and the VNC client runs on the computer. It is recommended to use **tightvncserver** on your Raspberry Pi. The steps are:

- Enter the following command:

  pi$raspberrypi:~ $ **sudo apt-get install tightvncserver**

- Run **tightvncserver**:

  pi$raspberrypi:~ $ **tightvncserver**

You will be prompted to create a password for remotely accessing the Raspberry Pi desktop. You can also set up an optional read-only password. The password should be entered every time you want to access the desktop. Enter a password and remember your password.

- Reboot the Raspberry Pi:

  pi@raspberrypi: ~ $ **sudo reboot**

- Start the VNC server by the following command:

  pi$raspberrypi:~ $ **vncserver :1**

  You can optionally specify screen size in pixels and colour depth in bits as follows:

  pi$raspberrypi:~ $ **vncserver :1 –geometry 1920x1080 –depth 24**

We must now set up a VNC viewer or client on our computer. There are many VNC clients available, but the recommended one which is compatible with **TightVNCServer** is **TightVNC** for the PC, which can be downloaded from the following link:

https://www.tightvnc.com/download.php

- Download and install the **TightVNC** software for your computer. You will have to choose a password during the installation.

- Start the **TightVNC Viewer** on your PC and enter the Raspberry Pi's IP address with ':1' appended to it. Click **Connect** (Figure 10.38)

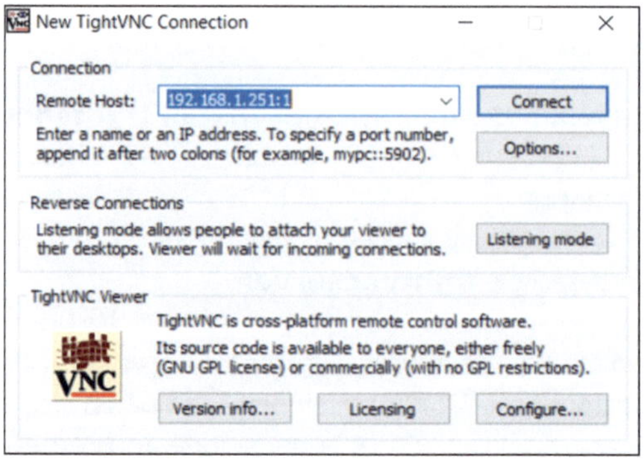

Figure 10.38 Connect with TightVNC viewer to the Raspberry Pi.

- Enter the password you have chosen earlier. You should now see the Raspberry Pi's desktop displayed on your computer screen (Figure 10.39)

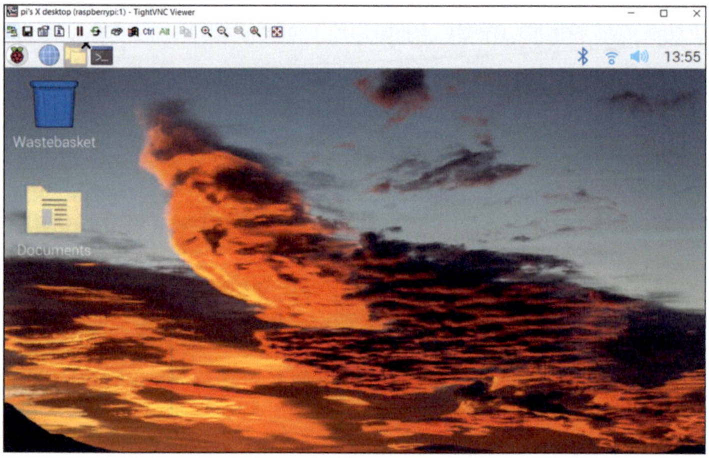

Figure 10.39 The desktop of the Raspberry Pi on your computer's screen.

You now have full access to the Raspberry Pi's desktop.

Raspberry Pi GPIO Pin Definitions

The Raspberry Pi has a 2×20 pin connector on which an extension board can be plugged, a so-called HAT (from 'Hardware Attached on Top'). This connector provides access to GPIO ports of the computer, and you can use them in your own projects. Figure 10.40 shows the pinout of the connector.

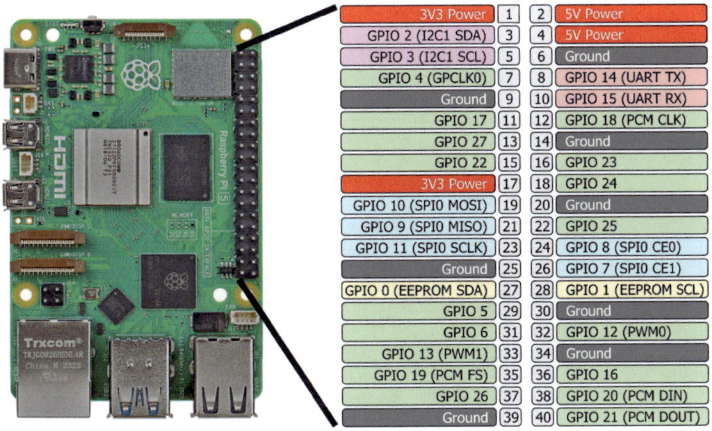

Figure 10.40 Extension connector pinout.

The connector provides 26 general-purpose bidirectional I/O pins. Some of the pins have multiple functions. For example, pins 3 and 5 are the GPIO2 and GPIO3 pins, respectively. These pins can also be used as an I²C bus with the SDA and SCL signals, respectively. Similarly, pins 19, 21 and 23 can either be used as GPIO10, GPIO9 and GPIO11 respectively, or as the SPI bus. Pins 8 and 10 are available for UART-type serial communication.

Two power outputs are provided: 3.3 V and 5.0 V. The GPIO ports operate with 3.3 V logic levels (unlike many other computer circuits that work with 5 V levels). A pin can either be an input or an output. When configured as an output, the pin voltage is either 0 V (logic 0) or 3.3 V (logic 1). When configured as an input, a voltage above 1.7 V will be considered a logic 1; a voltage below 1.7 V will be considered as a logic 0. Care should be taken not to supply voltages greater than 3.3 V to any I/O pin, as large voltages can easily damage your Raspberry Pi. The Raspberry Pi has no overvoltage protection circuitry.

The Raspberry Pi normally operates from an external power supply (e.g. a mains adapter) with a 5 V output. The amount of current that can be supplied by the 5 V pin depends on many factors such as the current required by the Raspberry Pi itself, the current taken by the USB peripherals, the camera current, the HDMI port currents, and so on. The 3.3 V output pin can supply up to 16 mA of current. The total current drawn at the same time from all output pins should not exceed the 51-mA limit. Care should be taken when connecting external devices to the GPIO pins, as drawing excessive currents or short-circuiting a pin can easily damage your Raspberry Pi.

10.6.2 Project 4 – Blinking an LED With a Raspberry Pi

Description: This is perhaps the easiest hardware project you can do with a Raspberry Pi. In this project, you will connect an LED to one of the GPIO ports of the Raspberry Pi and then toggle the state of the LED every second. The aim of this project is to show how a simple Python program can be written and then run from a file. The project also shows how to connect an LED to a GPIO pin of the Raspberry Pi. In addition, the project shows how to use the GPIO library to configure and set a GPIO pin to a logic 0 or 1.

Block diagram: The high-level overview of the project is shown in Figure 10.41.

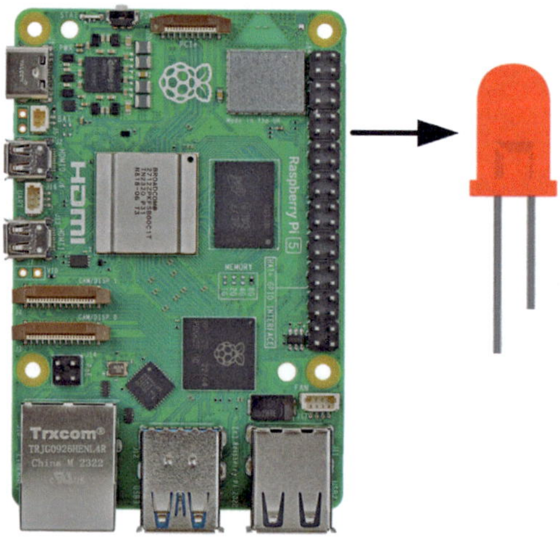

Figure 10.41 High-level overview of the project

Circuit diagram: The circuit diagram of the project is shown in Figure 10.42. An LED is connected to port pin GPIO 17 (pin 11) of the Raspberry Pi through a 470-Ω current-limiting resistor.

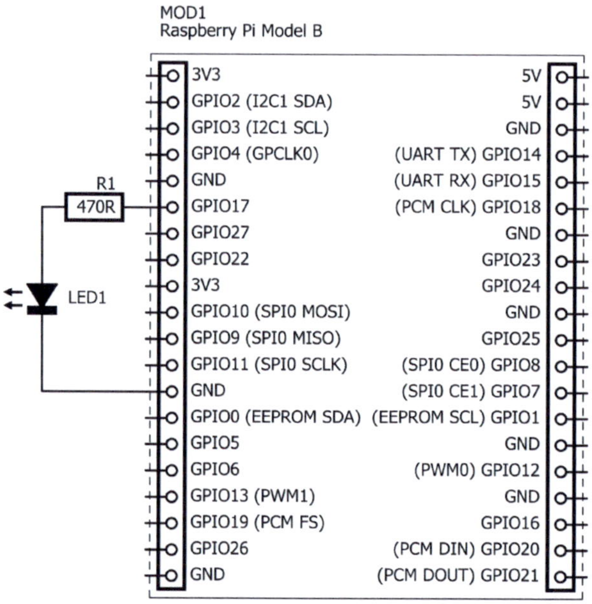

Figure 10.42 Circuit diagram of the project.

Construction: The project is constructed on a breadboard as shown in Figure 10.43. Jumper wires are used to connect the LED to the GPIO port. Notice that the short side of the LED must be connected to ground.

Figure 10.43 Constructing the project on a breadboard.

Program listing: The program is called **LED.py** and the listing is shown in Figure 10.44. The program was written using the **nano** text editor. At the beginning of the program, the **gpiozero** and the **time** modules are imported into the project. The rest of the program is executed indefinitely in a **while** loop, where the LED is turned ON and OFF with a one-second delay between each state change. Press Ctrl+C to terminate the program.

```
#--------------------------------------------------------
#
#                      FLASHING LED
#                      ============
#
# In thisproject a small LED is connected to GPIO 17 of
# the Raspberry Pi 5. The program flashes the LED every
# second.
#
# Program: LED.py
# Date   : December, 2023
# Author : Dogan Ibrahim
#---------------------------------------------------------
from gpiozero import LED        # import gpiozero
from time import sleep          # import time library

led = LED(17)

while True:
```

```
led.on()                          # turn ON LED
sleep(1)                          # wait 1 second
led.off()                         # turn OFF LED
sleep(1)                          # wait 1 second
```

Figure 10.44 Program listing of the project.

The program is run from the console as follows:

pi@raspberrypi ~ $ **python LED.py**

If you wish to run the program from the desktop, you should either connect a monitor to the Raspberry Pi or use the **VNC Viewer** as explained before. Then, click the **Applications menu → Programming → Thonny**. Click **File** and open file **LED.py**. You can also type the program in Thonny if you didn't create the file. Now, click **Run** to execute the program. You should see the LED change state every second. To terminate the program, close the screen by clicking the **STOP** button.

Chapter 11 • Other Topics of Interest

11.1 Overview

In this chapter, we will be looking at some other electronic topics that may be of interest to readers. These are topics such as component selection, reading a datasheet, PCB design, EMC and EMI, and electronic norms and regulations.

11.2 Component Selection

It is important to realize that poor component selection will often lead to problems in the design and the final product. Therefore, it is important to choose the components with care. Some guidelines are given here on how to choose the correct components:

- Use components that meet the circuit requirements. The components chosen must satisfy the voltage, current, and temperature ratings of the circuit to be constructed. For example, if the circuit operates from a 9-V power supply and uses electrolytic capacitors, you cannot use electrolytic capacitors rated at 5 V. Similarly, if a resistor carries high current, then it must be selected to have the correct wattage. A transistor that may get hot in a circuit must be cooled with, for example, a heat sink or a fan.

- When designing a circuit, make sure that the components you choose can be bought easily. For example, it is not good design practice to specify a resistor with a non-standard value of 4.713 Ω. Choose the nearest value that will not change the behaviour of the circuit. In this case, specify 4.7 Ω instead.

- Pay attention to the availability of parts. Do not use parts that are not available and that can take weeks or months to obtain them. Avoid using obsolete components in your design. Such components may not be available and may have been replaced with more up-to-date ones. Always use newer components wherever possible. Check the manufacturer's website for information about a component's status (in production, not recommended for new designs, obsolete, replaced, etc.).

- If you are designing or publishing an electronic circuit, use internationally accepted component symbols.

- Minimum order quantity is another point to be considered. Some suppliers send out multiple units, even though you may want only a few. For example, you may want a couple of diodes, but the suppliers may send out a minimum quantity of 100 diodes at a time. If this is the case, it may be worthwhile to recommend vendors where the readers can purchase the components without a minimum order quantity.

- Do not use unnecessary components as this may increase the cost of the project. For example, if you want to lower the supply voltage from 9 V to 5 V, use a resistive potential circuit if possible, or a cheap Zener diode instead of designing a full-blown power supply. The acronym KISS (Keep It Simple, Son) is the thing to keep in mind.

- nclude environmental factors when designing and building your projects. For example, if the project is to be used at the North Pole, operation and storage of the project at very low temperatures must be considered.

- Consider the costs. If possible, specify and use lower-cost components to save money on the project.

This list of best practices should always be kept in mind, but it is not a law. You are free to take liberties and adapt them when needed. The main goal is making your project a success.

11.3 Reading a Datasheet

Most component manufacturers and suppliers provide free datasheets of the components they offer. It is important to go through these datasheets and make sure that the component to be used satisfies the requirements. For example, if you are designing a 9 V battery-operated project and one of the components you have chosen without reading the datasheet requires a minimum of 12 V, then you have a problem as this component will not function properly.

As an example datasheet, we will look at the datasheet of temperature sensor LM35, manufactured by Texas Instruments. This datasheet is freely available on the Internet. Figure 11.1 shows the first page of the datasheet. Here we see the name of the component at the top. At the left-hand side are the features of the component, and below it are some typical applications. Looking at the features, we can immediately see the important features of this component, such as its linearity, temperature range, accuracy, power supply range, etc. If we are happy with the features, we can carry on looking at more details on the component. On the right-hand side, we see a brief description of the component and its package types.

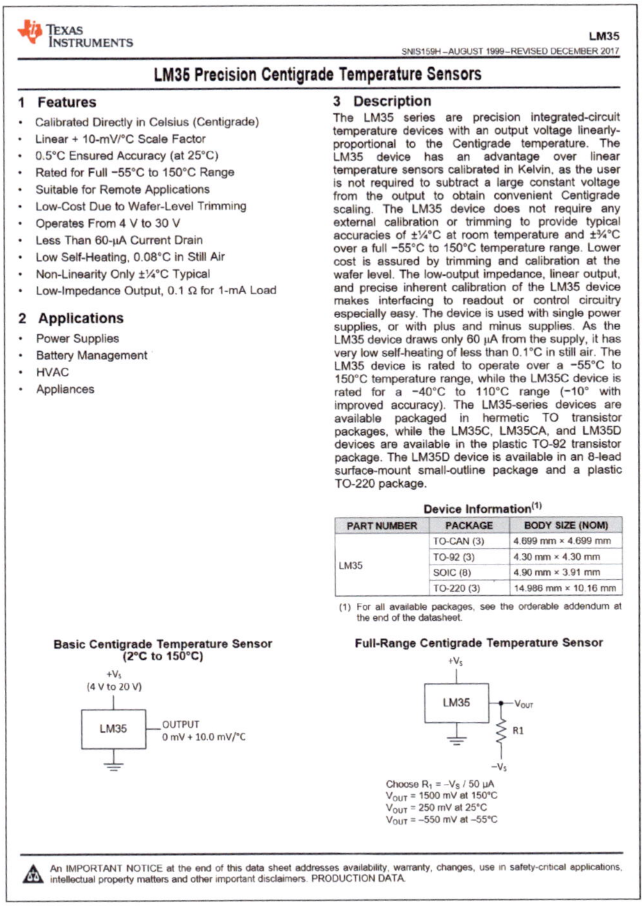

TEXAS INSTRUMENTS

LM35

SNIS159H – AUGUST 1999 – REVISED DECEMBER 2017

LM35 Precision Centigrade Temperature Sensors

1 Features

- Calibrated Directly in Celsius (Centigrade)
- Linear + 10-mV/°C Scale Factor
- 0.5°C Ensured Accuracy (at 25°C)
- Rated for Full −55°C to 150°C Range
- Suitable for Remote Applications
- Low-Cost Due to Wafer-Level Trimming
- Operates From 4 V to 30 V
- Less Than 60-μA Current Drain
- Low Self-Heating, 0.08°C in Still Air
- Non-Linearity Only ±¼°C Typical
- Low-Impedance Output, 0.1 Ω for 1-mA Load

2 Applications

- Power Supplies
- Battery Management
- HVAC
- Appliances

3 Description

The LM35 series are precision integrated-circuit temperature devices with an output voltage linearly-proportional to the Centigrade temperature. The LM35 device has an advantage over linear temperature sensors calibrated in Kelvin, as the user is not required to subtract a large constant voltage from the output to obtain convenient Centigrade scaling. The LM35 device does not require any external calibration or trimming to provide typical accuracies of ±¼°C at room temperature and ±¾°C over a full −55°C to 150°C temperature range. Lower cost is assured by trimming and calibration at the wafer level. The low-output impedance, linear output, and precise inherent calibration of the LM35 device makes interfacing to readout or control circuitry especially easy. The device is used with single power supplies, or with plus and minus supplies. As the LM35 device draws only 60 μA from the supply, it has very low self-heating of less than 0.1°C in still air. The LM35 device is rated to operate over a −55°C to 150°C temperature range, while the LM35C device is rated for a −40°C to 110°C range (−10° with improved accuracy). The LM35-series devices are available packaged in hermetic TO transistor packages, while the LM35C, LM35CA, and LM35D devices are available in the plastic TO-92 transistor package. The LM35D device is available in an 8-lead surface-mount small-outline package and a plastic TO-220 package.

Device Information(1)

PART NUMBER	PACKAGE	BODY SIZE (NOM)
LM35	TO-CAN (3)	4.699 mm × 4.699 mm
	TO-92 (3)	4.30 mm × 4.30 mm
	SOIC (8)	4.90 mm × 3.91 mm
	TO-220 (3)	14.986 mm × 10.16 mm

(1) For all available packages, see the orderable addendum at the end of the datasheet.

Basic Centigrade Temperature Sensor (2°C to 150°C)

Full-Range Centigrade Temperature Sensor

An IMPORTANT NOTICE at the end of this data sheet addresses availability, warranty, changes, use in safety-critical applications, intellectual property matters and other important disclaimers. PRODUCTION DATA.

Figure 11.1 First page of the datasheet of the LM35 (source: Texas Instruments).

Moving a few pages forward, after the table of contents, we find the pin configurations of the device for different packages (Figure 11.2).

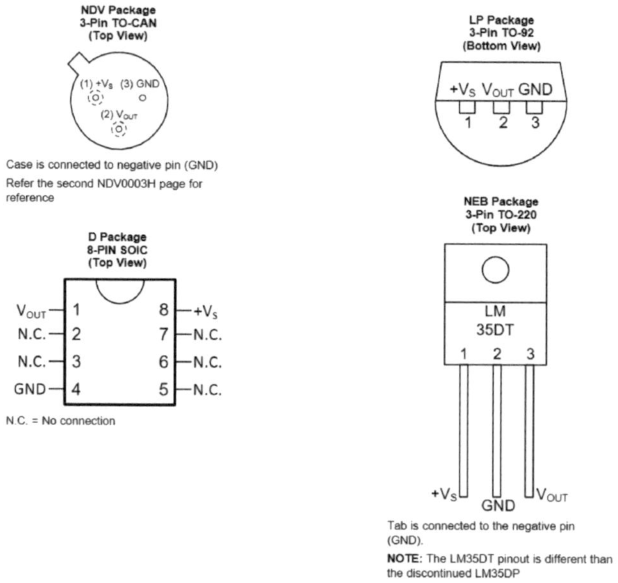

Figure 11.2 LM35 packages and pin configurations (source: Texas Instruments).

Again, browsing forward, we discover the Absolute Maximum Ratings (Figure 11.3). The information given here is essential because if we operate the device outside the limits given here, the device may malfunction or even be damaged. For example, the maximum supply voltage is given as 35 V, and it is recommended not to operate the device at or close to this voltage.

6 Specifications

6.1 Absolute Maximum Ratings

over operating free-air temperature range (unless otherwise noted)(1)(2)

		MIN	MAX	UNIT
Supply voltage		–0.2	35	V
Output voltage		–1	6	V
Output current			10	mA
Maximum Junction Temperature, $T_J$max			150	°C
Storage Temperature, T_{stg}	TO-CAN, TO-92 Package	–60	150	°C
	TO-220, SOIC Package	–65	150	

(1) If Military/Aerospace specified devices are required, please contact the Texas Instruments Sales Office/ Distributors for availability and specifications.
(2) Absolute Maximum Ratings indicate limits beyond which damage to the device may occur. DC and AC electrical specifications do not apply when operating the device beyond its rated operating conditions.

6.2 ESD Ratings

			VALUE	UNIT
$V_{(ESD)}$	Electrostatic discharge	Human-body model (HBM), per ANSI/ESDA/JEDEC JS-001(1)	±2500	V

(1) JEDEC document JEP155 states that 500-V HBM allows safe manufacturing with a standard ESD control process.

6.3 Recommended Operating Conditions

over operating free-air temperature range (unless otherwise noted)

		MIN	MAX	UNIT
Specified operating temperature: T_{MIN} to T_{MAX}	LM35, LM35A	–55	150	°C
	LM35C, LM35CA	–40	110	
	LM35D	0	100	
Supply Voltage (+V_S)		4	30	V

Figure 11.3 Never ignore the Absolute Maximum Ratings (source: Texas Instruments).

Then we come to the Electrical Characteristics. Here, the typical, tested and design limits of all parameters of interest to the device are given for each model of the device. Figure 11.4 shows a typical Electrical Characteristics page.

6.5 Electrical Characteristics: LM35A, LM35CA Limits

Unless otherwise noted, these specifications apply: $-55°C \le T_J \le 150°C$ for the LM35 and LM35A; $-40°C \le T_J \le 110°C$ for the LM35C and LM35CA; and $0°C \le T_J \le 100°C$ for the LM35D. V_S = 5 Vdc and I_{LOAD} = 50 µA, in the circuit of Full-Range Centigrade Temperature Sensor. These specifications also apply from 2°C to T_{MAX} in the circuit of Figure 14.

PARAMETER	TEST CONDITIONS	LM35A TYP	LM35A TESTED LIMIT(1)	LM35A DESIGN LIMIT(2)	LM35CA TYP	LM35CA TESTED LIMIT(1)	LM35CA DESIGN LIMIT(2)	UNIT
Accuracy(3)	T_A = 25°C	±0.2	±0.5		±0.2	±0.5		°C
	T_A = –10°C	±0.3			±0.3		±1	
	$T_A = T_{MAX}$	±0.4	±1		±0.4	±1		
	$T_A = T_{MIN}$	±0.4	±1		±0.4		±1.5	
Nonlinearity(4)	$T_{MIN} \le T_A \le T_{MAX}$, $-40°C \le T_J \le 125°C$	±0.18		±0.35	±0.15		±0.3	°C
Sensor gain (average slope)	$T_{MIN} \le T_A \le T_{MAX}$	10	9.9		10		9.9	mV/°C
	$-40°C \le T_J \le 125°C$	10	10.1		10		10.1	
Load regulation(5) $0 \le I_L \le 1$ mA	T_A = 25°C	±0.4	±1		±0.4	±1		mV/mA
	$T_{MIN} \le T_A \le T_{MAX}$, $-40°C \le T_J \le 125°C$	±0.5		±3	±0.5		±3	
Line regulation(5)	T_A = 25°C	±0.01	±0.05		±0.01	±0.05		mV/V
	$4 V \le V_S \le 30 V$, $-40°C \le T_J \le 125°C$	±0.02		±0.1	±0.02		±0.1	
Quiescent current(6)	V_S = 5 V, 25°C	56	67		56	67		µA
	V_S = 5 V, $-40°C \le T_J \le 125°C$	105		131	91		114	
	V_S = 30 V, 25°C	56.2	68		56.2	68		
	V_S = 30 V, $-40°C \le T_J \le 125°C$	105.5		133	91.5		116	
Change of quiescent current(5)	$4 V \le V_S \le 30 V$, 25°C	0.2	1		0.2	1		µA
	$4 V \le V_S \le 30 V$, $-40°C \le T_J \le 125°C$	0.5		2	0.5		2	
Temperature coefficient of quiescent current	$-40°C \le T_J \le 125°C$	0.39		0.5	0.39		0.5	µA/°C
Minimum temperature for rate accuracy	In circuit of Figure 14, I_L = 0	1.5		2	1.5		2	°C
Long term stability	$T_J = T_{MAX}$, for 1000 hours	±0.08			±0.08			°C

(1) Tested Limits are ensured and 100% tested in production.
(2) Design Limits are ensured (but not 100% production tested) over the indicated temperature and supply voltage ranges. These limits are not used to calculate outgoing quality levels.
(3) Accuracy is defined as the error between the output voltage and 10 mv/°C times the case temperature of the device, at specified conditions of voltage, current, and temperature (expressed in °C).
(4) Non-linearity is defined as the deviation of the output-voltage-versus-temperature curve from the best-fit straight line, over the rated temperature range of the device.
(5) Regulation is measured at constant junction temperature, using pulse testing with a low duty cycle. Changes in output due to heating effects can be computed by multiplying the internal dissipation by the thermal resistance.
(6) Quiescent current is defined in the circuit of Figure 14.

Figure 11.4 Electrical Characteristics of the LM35 (source: Texas Instruments).

Browsing further, we find the Typical Characteristics in the form of graphs. Here, the graphs show the thermal resistance, accuracy vs. temperature, noise voltage, etc. (Figure 11.5)

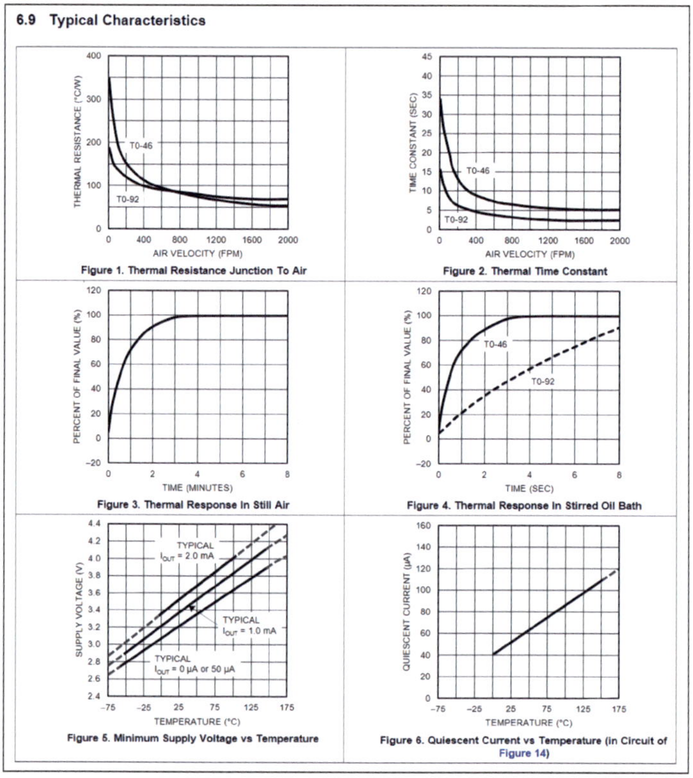

Figure 11.5 Typical Characteristics presented as graphs (source: Texas Instruments).

The remainder of the datasheet gives detailed descriptions about the component, example application circuits, and packaging and ordering information, etc.

Sometimes datasheets are very detailed, sometimes they are (too) brief. Sometimes they get revised, so make sure you always have the most recent version.

11.4 Schematic Capture and PCB Design

Schematic capture is the process of drawing an electronic circuit using a software program. We have seen in examples in earlier chapters how to use the TINA-TI simulator to draw (and simulate) electronic circuits.

Some best-practice rules for schematic capture are:

- All component symbols should comply with the standardized electronic symbols. Signals flow from left to right. This means that components should be placed such that the left sides are the inputs, and the right sides are the outputs.

- Power supplies flow from top to bottom. The highest positive supply voltage is at the top of the drawing, the lowest negative supply voltage is at the bottom. Power supply and ground connections must be clearly marked.

- Wires should not cross other wires if possible. A connection of two or more wires should be marked with a junction dot to make them clearly visible. Avoid 4-wire junctions; use two 3-wire junctions instead (as an example, see Figure 5.19, the junction of signal A, T1 and C1).
- Labels, names, and references must be consistent throughout the drawing. Component names or part numbers should be included in the drawings. For example, resistors should be marked as R1, R2, R3 etc., and their values should be included in the drawing if possible.
- If the drawing is too large to fit on a single page, use multiple pages, where each page contains a complete subcircuit. Using hierarchical sheets is a good way to keep large schematics readable.
- Add notes and comments to clarify points which may not be speaking for themselves. Such remarks can concern a particular placement requirement on the PCB, cooling, etc.

The output of the schematic capture is generally in the form of a **netlist** that lists all the components and their connections. An example circuit and its netlist are shown in Figure 11.6 and Figure 11.7 respectively.

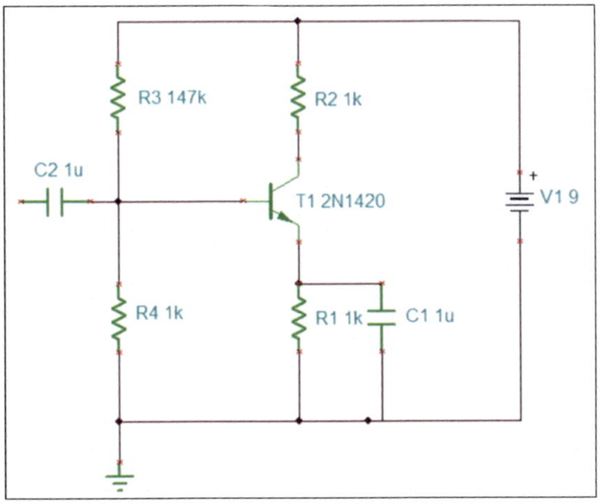

Figure 11.6 Example circuit drawn in TINA-TI.

```
Common-emitter (TINA Netlist Editor format)
*************************************
**  This file was created by TINA   **
**        www.tina.com               **
**      (c) DesignSoft, Inc.         **
**     www.designsoftware.com        **
*************************************
```

```
.TEMP 27
.TRAN 1P 1U

.OPTIONS NOSTEPDEP

V1          1 0 9
C2          2 3  1U
R4          3 0 1K
R3          1 3 147K
R2          1 4 1K
C1          5 0  1U
R1          5 0 1K
QT1         4 3 5  Q_2N1420_N_1

.MODEL Q_2N1420_N_1 NPN( IS=40.1F NF=1 NR=1 RE=834M RC=834M
+       RB=33.3 VAF=73.3 VAR=36.7 ISE=51F ISC=51F
+       ISS=0 NE=1.42 NC=1.42 NS=1 BF=170
+       BR=5 IKF=366M IKR=366M CJC=9.28P CJE=22.1P
+       CJS=0 VJC=301M VJE=657M VJS=750M MJC=247M
+       MJE=351M MJS=0 TF=3.18N TR=414N EG=1.11
+       KF=0 AF=1 )


.END
```

Figure 11.7 Netlist of the circuit of Figure 11.6.

The netlist is usually imported into a PCB (Printed Circuit Board) design program so that the circuit can be built on a PCB. There are several free PCB design packages available on the Internet. KiCad is one of them. The author uses the **TINA Design Suite** for schematic capture, circuit simulation, and PCB design. This is a full-featured professional tool, and it is not free. Although it has some similarities to TINA-TI, it includes many more tools. Figure 11.8 shows the PCB layout of the circuit designed in Figure 11.6. This board has been 'auto-routed' so that all the connections are made automatically by the design tool.

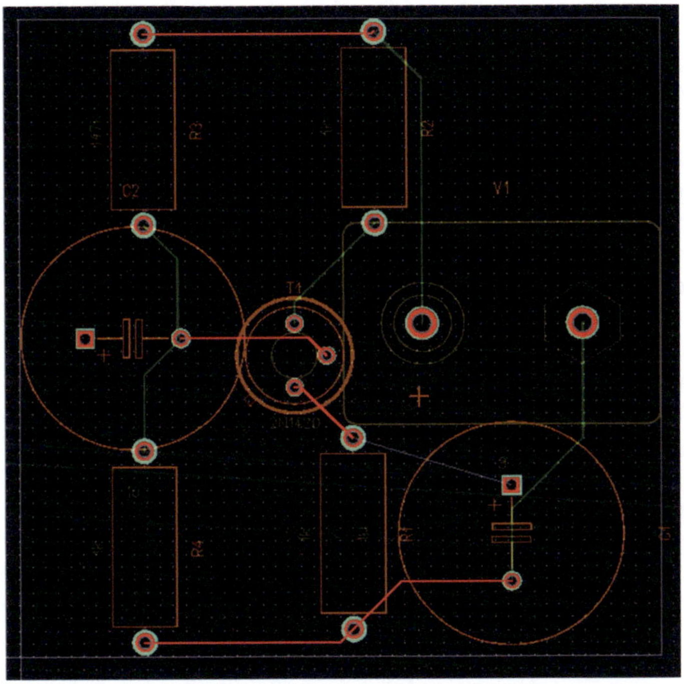

Figure 11.8 PCB layout of the circuit in Figure 11.6.

When the PCB layout is ready, we must export the manufacturing files. These are usually (but not always) files in gerber and excellon format. The gerber files contain the drawing itself, the excellon (drill) files specify the hole diameters and positions. They are used by the PCB manufacturer to make the PCB according to your specifications. If the board manufacturer must also mount the components, you should provide a pick-and-place file that specifies the position of the components on the board.

The stages to design and produce an electronic product that you can sell are summarized below:

1. Design the circuit using pen and paper. Start at a high overview level and work your way down to the details. Think about what it should look like and how it is to be used.

2. Simulate the circuit. Once you are happy with it, continue to the next step.

3. Build a prototype on a breadboard or on a perforated prototyping card. Make sure it works as intended before continuing. Also, ensure that the components can be bought easily.

4. Draw the circuit diagram using a suitable computer program.

5. Import the netlist into a PCB design tool. Define a board size and shape and place the components on the board. Depending on the tool you are using, component placement can either be done manually or automatically.

6. Guided by the imported netlist and the "rat's nest", draw the traces to connect the components. Depending on the tool you are using, this can either be done manually or automatically. Respect the PCB design rules (DRC, they concern trace widths, isolation distances, minimum hole size, etc.) to avoid problems later during manufacturing.

7. Export the board manufacturing files (gerber and drill) and send them to a PCB manufacturer.

8. Populate the designed PCB with the components and test it. If you are satisfied with the design, then continue. More likely though is that you will have to go back to a previous step because you discovered a mistake.

9. Finalize the assembly with a suitable container that you specified during the first step. This will probably require going back to a previous step to change or fix some detail that you didn't foresee earlier.

10. Produce many copies of the product, tell the world about it, and start selling.

11.5 EMC and EMI

The term EMC stands for Electromagnetic Compatibility, and EMI stands for Electromagnetic Interference.

EMC Testing

A sensitive electronic device can be affected by electromagnetic radiation present in its environment. Nearby LED lighting, electric motors and other heavy electrical equipment can create electromagnetic disturbances, and these may affect the functioning of other electronic devices. EMC testing ensures that an electronic device is immune to electromagnetic interference around it and that it doesn't produce interference that may hinder other equipment.

For a product to be EMC compliant, the correct components must be selected, and it must be constructed properly during the design, development and production stages. Unless EMC tests are carried out, the final product may not be safe, may fail, or may not function as intended in certain situations.

EMC testing is important in any product design. It requires the use of specialised EMC test and monitoring equipment, such as amplifiers, receiving antennas, special receivers, spectrum analyzers, and electromagnetic sensors. Accurate measurements are carried out at various frequencies to measure the noise generated by the device and to see if and how it is influenced by electromagnetic interference. EMC testing is usually outsourced to a specialised lab.

Many countries have regulatory guidelines for EMC testing. CE marking has been required for the sale of almost all products within the European Union (EU) since 1985. The CE mark indicates that the product meets required standards and protection of the environment requirements, and therefore that it can move within the European Union without any restrictions. If you want to sell a component such as a transistor or a resistor, you don't have to stick a CE label on it because these are not considered to be equipment. However, if you manufacture radios, TV sets, or washing machines, and you wish to sell them in the EU, then you must obtain the correct CE certificates and add CE markings on visible parts of these products. For selling something in the USA, you will need to do all this as well for the FCC.

EMI

Electromagnetic interference can occur in many ways and arises from various sources. It can either be man-made, or it can naturally occur in the environment. Man-made EMI usually emits from electronic circuits and from switching heavy electrical equipment such as motors. Naturally occurring EMI arises from sources in the environment such as cosmic rays, radiation, lightning, and other environmental sources. EMI can be transferred by conduction (e.g. through wires) or by radiation. The noise generated by electrical and electronic devices can propagate along power lines and other wire connections, and this can affect other devices which are in contact with it. Radiated interference occurs from sources without any physical contact. This form of EMI is propagated through the air. An example of radiated interference is the interference caused by heavy motors and pumps or a strong, nearby radio station.

The EMI generated by an electronic device can be reduced in many ways. Grounding or shielding wires can reduce the EMI considerably. Ferrite beads or chokes are used in series with or around input and output wires to suppress radiation. Shielded wires and enclosures not only reduce the interference generated by the device, they also help to improve its immunity against interference from other nearby electrical and electronic sources.

11.6 Norms and Regulations

Every country has a set of standards and norms which specify criteria that (electronic) equipment must comply with. These criteria are intended to ensure user-and environmental safety. They concern the materials used, the way the equipment is assembled, how it can be used and in which situations, what the maximum stress levels for safe operation are, how testing should be done, etc., etc. Most products distributed in the European Union are subject to these safety standards and testing requirements.

Although much of these standards are for the manufacturers to guide them what types of materials to use, etc., the regulations include the consumers as well. For example, any device which includes mercury must be disposed of according to the regulations. One cannot simply dispose of such a device by putting it in the bin.

Manufacturers and importers of electrical and electronic equipment should know and follow the RoHS directive (**R**estriction **o**f the use of certain **H**azardous **S**ubstances in electrical and electronic equipment) and REACH regulations (**R**egistration, **E**valuation, **A**uthorization and Restriction of **CH**emicals). The details of these regulations are available in detail on the Internet. For example, the RoHS directives restrict the use of hazardous substances such as heavy metals in electrical and electronic equipment. This directive applies to electrical equipment such as household appliances, electrical and electronic tools, lighting equipment, and general home equipment.

It is important to make sure that every component used in a product is RoHS compliant. If a single component is not compliant, the whole electronic product is assumed to be non-compliant. The problem is that not all component manufacturers provide RoHS test reports. In such cases, you have to consider third party RoHS compliance test centers to have your components tested and certified. There is a list of measures that must be taken by the manufacturers for the safety of electrical and electronic devices. RoHS includes many directives with the aims to:

- Use lead-free solder.
- Limit the use of toxic substances such as lead, mercury, hexavalent chromium, cadmium and several others.
- Manufacture electrical and electronic components that are not hazardous.
- Ban toxic flame retardants in the manufacture of cable insulation.
- Avoid the use of sharp objects in electrical and electronic equipment.

Appendix – List of Components Used in Projects

- 39 Ω, 1 W resistor
- 47 Ω resistor
- 180 Ω resistor
- 330 Ω resistor
- 3× 1 kΩ resistor
- 2.2 kΩ resistor
- 3.9 kΩ resistor
- 6.8 kΩ resistor
- 10 kΩ resistor
- 15 kΩ resistor
- 22 kΩ resistor
- 33 kΩ resistor
- 47 kΩ resistor
- 56 kΩ resistor
- 82 kΩ resistor
- 120 kΩ resistor
- 680 kΩ resistor
- 2× 100 kΩ resistor
- 10 kΩ trimmer
- 10 kΩ potentiometer
- 100 kΩ potentiometer
- Light dependent resistor (LDR)
- 1 nF ceramic capacitor
- 2× 10 nF ceramic capacitor
- 100 nF ceramic capacitor
- 1 µF, 25 V electrolytic capacitor
- 2× 10 µF, 25 V electrolytic capacitor
- 100 µF, 25 V electrolytic capacitor
- 470 µF, 25 V electrolytic capacitor
- 1000 µF, 25 V electrolytic capacitor
- Common-cathode RGB LED
- 1N4148 diode
- 5.1 V, 1 W Zener diode
- 3× 5 mm red LED
- 2× BC337 NPN transistor
- IRFZ44N N-channel MOSFET
- 2× 555 timer IC
- LM393 comparator IC
- 74HCT08 AND gate
- 3× Push-button switch
- 2× SPDT switch
- 9 V SPDT relay
- Active buzzer
- Passive buzzer
- Solid wire
- 2× 9 V PP-3 battery clip with wires
- Breadboard
- Jumper wires, male-to-male
- 2× 9 V PP-3 battery

Index